Encyclopedia of
Sustainability

Encyclopedia of
Sustainability

Environment and Ecology
VOLUME I

Robin Morris Collin

Robert William Collin

GREENWOOD PRESS
An Imprint of ABC-CLIO, LLC

A B C 🔻 C L I O

Santa Barbara, California • Denver, Colorado • Oxford, England

Library of Congress Cataloging-in-Publication Data

Encyclopedia of sustainability / Robin Morris Collin, Robert William Collin.
   3 v. cm.
  Includes bibliographical references and index.
  Contents: vol. 1. Environment and ecology —
  ISBN 978-0-313-35263-8 (vol. 1 print : alk. paper) — ISBN 978-0-313-35264-5 (vol. 1 e-book) — ISBN 978-0-313-35265-2 (vol. 2 print : alk. paper) — ISBN 978-0-313-35266-9 (vol. 2 e-book) — ISBN 978-0-313-35267-6 (vol. 3 print : alk. paper) — ISBN 978-0-313-35268-3 (vol. 3 e-book) — ISBN 978-0-313-35261-4 (set - print : alk. paper) — ISBN 978-0-313-35262-1 (set - e-book)
  1. Environmental sciences—Encyclopedias. 2. Sustainability—Encyclopedias. 3. Sustainable development—Encyclopedias. I. Collin, Robin Morris. II. Collin, Robert W., 1957–
  GE10.E528 2010
  333.7203—dc22   2009037029

14 13 12 11 10   1 2 3 4 5

This book is also available on the World Wide Web as an eBook.
Visit www.abc-clio.com for details.

ABC-CLIO, LLC
130 Cremona Drive, P.O. Box 1911
Santa Barbara, California 93116-1911

This book is printed on acid-free paper ∞

Manufactured in the United States of America

CONTENTS

GUIDE TO RELATED TOPICS

PREFACE

References to sustainability are everywhere, from advertising to space travel. Words associated with sustainability are fast becoming ubiquitous. This reference text is designed to help understand the many meanings of sustainability.

Concepts of sustainability have been developed in multiple disciplines including the sciences, international agreements, development law and policy, and humanities. The essential concepts about sustainability can be described in terms of three broad domains:

- Environment and ecology

- Business and economics

- Equity and fairness

The relationship between these domains is described in somewhat different ways. Some describe their relationship as a three-legged stool or three intersecting circles. Each circle or leg of the stool represents one domain, environment, economics, and equity. Each circle is of equal size; each leg bears equal weight.

Others describe the fundamental relationships as three nested baskets. The environment is the largest, most comprehensive basket. Within it, all human activity is located including human economic enterprises and human communities. Economic enterprise is represented by a basket nested within our environment and its webs of life. Human

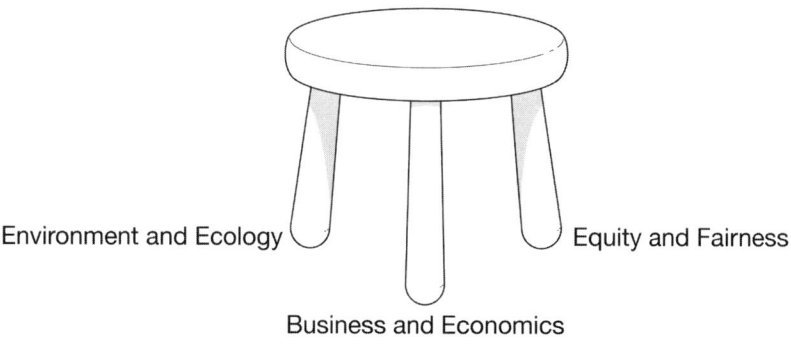

Figure 1.1 • Sustainability as a three-legged stool: environment, economics, and equity. Illustrator: Jeff Dixon.

individuals and their communities rest within both these two baskets relying on each for their livelihood and support. This encyclopedia set will provide the reader with the necessary infrastructure to navigate the complex roads and byways of the contemporary discourse on sustainability. That infrastructure is based on the axes of environment and ecology, economics and business, and equity or fairness. Sustainability weights these three areas equally and joins them in every discourse. This triangulation of the so-called three "E"s—environment, economics, and equity—distinguishes sustainability as a philosophy different from that of conservationism or environmentalism. This encyclopedia devotes one volume to each of the three "E"s.

In each volume, there is the same basic organization of chapters: a comprehensive introduction, definitions in contexts that are pertinent to that particular volume, the contemporary public policy contexts arranged from the global to national to local levels, current controversies, and future trends.

Each volume begins with a comprehensive overview of what the term *sustainability* means in each domain. This comprehensive overview introduces the major concepts of sustainability as used in each unique

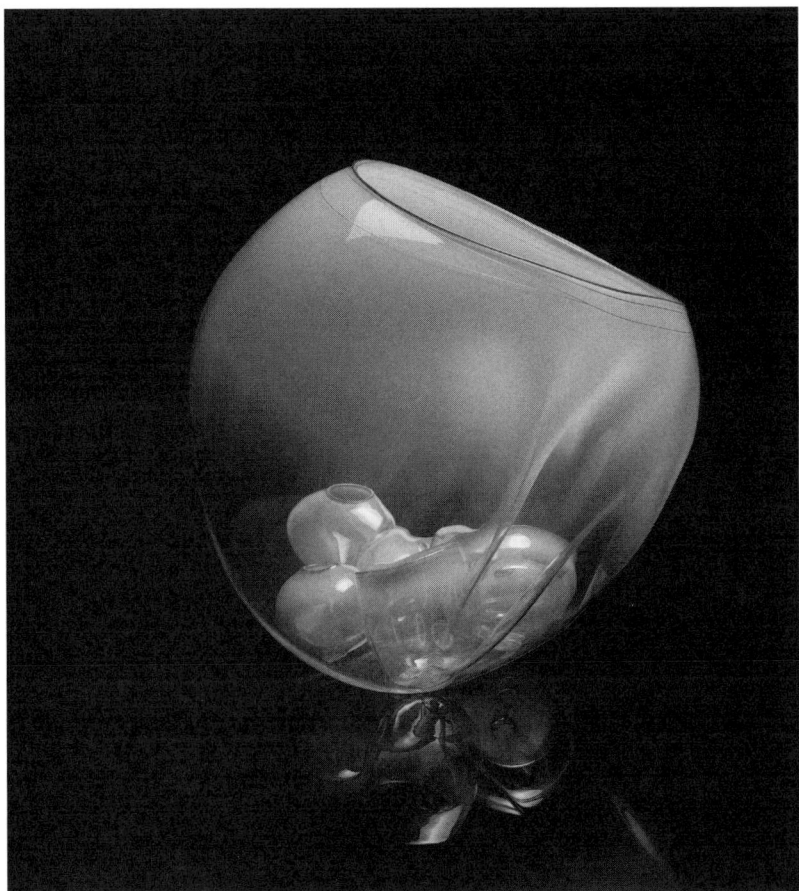

FIGURE 1.2 • Sustainability as three nested baskets: Environment, economics, and equity. Copyright © Dale Chihuly, 1992. Sky Blue Basket Set with Cobalt Lip Wraps, 1992 17″ × 15″ × 16″. Photo by Terry Rishel.

volume: environment and ecology, economics and business, and equity or fairness. This introductory chapter provides a concentrated account of the "big picture" in each unique arena of sustainability. The interconnectedness of basic ideas makes the study of sustainability challenging to the novice and complex to anyone. The overview section of each volume presents basic concepts and relationships in the primary context of the volume. The overview provides concise insight into terms of art, the dynamics that have shaped the concept within that context, and the changes and challenges that sustainability presents in that particular context. These are the elements of complex interactions and the background on which human choices and policies act and interact with natural systems.

Following the overview, each volume contains a chapter on definitions and contexts that give in-depth descriptions of key terms set into contexts relevant to that particular area. This chapter is a primer on the basic terms and definitions that are foundational to each volume. Without such a primer, terms of art can become a private language of expertise, making knowledge and information inaccessible to the general public reader, even one with considerable education. This section identifies key terms of art and defines them in accessible language. Definitions are arranged contextually, and alphabetically. This organization is tailored to assisting the reader with quick and ready access to the language and contexts of the sustainability discourse. This chapter gives a reader ready access to the specialized terms of art and foundational definitions of the discourse in each area. The dynamic interaction of these terms is illustrated in images and examples throughout this volume.

Next in each volume Chapter 3 describes the role of government and the United Nations in achieving sustainability. We include a description of United Nations programs while recognizing that the UN is not a government but an association of sovereign governments. The UN has exercised global leadership in guiding world governments toward sustainability. We also describe the work of national, regional, and local governments. Local governments have a uniquely important role to play in implementing sustainability because they are most closely connected to place and community. Even in countries whose national governments have chosen to abdicate their role in achieving sustainability, local governments have acted independently as laboratories and as activist organizations to achieve important behavioral changes. Many United Nations programs and policies are aimed at this local and regional level of government as well as through non-governmental organizations (NGOs).

In the government section of each book, public policy is arranged in a global to local progression and, within that structure, it is presented in its chronological order. We have organized each chapter on government involvement to trace the historical developments as they have occurred at different levels: global developments through the United Nations organization, U.S. national developments where they have contributed, and the state, local and regional efforts within the United States. Sustainability public policy is rapidly unfolding in these different venues at uneven rates of change. For example, U.S. municipalities are rapidly adopting the Precautionary Principle in land use ordinances designed around sustainability.

In Chapter 4 of each volume, controversies related to sustainability are explored in greater depth. Every environmental and developmental conflict reflects competing and sometimes conflicting interests of many interested parties. Too often, these are portrayed in the popular media as simple conflicts between two parties. The truth about controversies, however, requires an appreciation of the competing and conflicting interests of multiple stakeholders. In the end, all solutions will be local, as are all environmental and ecological controversies. Solutions are not the focus of this section of the volumes here. Instead, we aim to describe the nature of the interests involved in all aspects. Controversies are once again arranged alphabetically. Each volume lists and describes current controversies in each volume. These controversies were selected based upon their political salience, and likely impact upon future generations. Controversies around sustainability provide a rich area for classroom discourse.

Finally, each volume concludes with a section devoted to emerging trends arranged alphabetically. This chapter takes a considered look at trends and data to follow for future developments. These trends were selected partially based upon the availability of existing data collections, and the commitment of governmental organizations and non-governmental organizations to collect and monitor data. New sources of data and better resources will continue to develop but these fundamental trends should lead the interested reader to the sources that we have now. In this section, we describe what the future of sustainability in each area may hold based on the information available now. At this moment, some changes seem inevitable, whereas others may be subject to human management. The future of our relationships built on a sustainable model of dynamic and radical inclusion is the subject of imagination and possibilities.

The concept of sustainability is divided into three co-equal components of environment, economy and equity. We did so to arrange the large amount of information in a manner most comprehensible to the reader. These components of sustainability are dynamic as well as interrelated. We developed our framework of overview, definitions, government involvement, controversies, and future trends per volume to help the reader understand sustainability in the context of each of the co-equal components. Within each framework, we have further introduced concepts of scale, such as going from global to local levels of government intervention. When appropriate, we have introduced chronologies that underscore the development of sustainability. Each volume is unique, and capable of standing alone, but with a similar framework. Each volume is cross-referenced, with portal websites. Our goal is to provide a comprehensive framework that the reader can easily navigate within and between volumes.

The ultimate challenge of human sustainability is how human enterprise and communities that can establish themselves without undermining the fundamental health of the natural systems on which all beings on Earth rely. The legacy of previous centuries, with their development of human enterprises and communities of great scale reliant on diminishing ecological resources, is the presence of wealth amidst poverty and environmental degradation that challenges our ability to survive. Sustainability as a doctrine challenges us to provide for our needs while allowing future generations the same opportunities for prosperity and a full experience of life. The idea that we can do that for future generations while ignoring the growing inequities of contemporary life is an equal challenge to sustainability. The central challenge of sustainability is how human enterprises and communities can function within ecosystems supporting our environment. The great human progress and development of the contemporary era have been accompanied by a growing gap between rich and poor people in the context of widespread environmental deterioration, and increasing poverty. Earth's ecosystems provide enormous benefits for humankind. Some of those benefits are from renewable resources, and some are not renewable. Humankind has exceeded the limits of some renewable resources and is approaching the limits of nonrenewable ones.

For those who are interested in the idea of sustainability and wish to explore it further, there is an overwhelming volume of material to read devoted to specific contexts and applications. Often, this material is difficult to penetrate for a novice because it is so heavily reliant on specialized language and specialized constructs unique to a particular discipline. Theses volumes provide a gateway that allows access to the full variety of the field and facilitates independent investigation in a multidisciplinary field. Sustainability will require new ways of thinking about the environment and a basic shift in public policy at all levels of government. This reference is dedicated to the task of facilitating human imagination and thought in that direction. Imagination is a uniquely human faculty praised in physics and metaphysics alike. Albert Einstein said that imagination was more important than information. Buddhism insists that thought and intention are as important as the acts to which they may give birth. Creativity may be inspired by the interplay between the major perspectives offered in each volume. Pragmatic implementation is illustrated in stories, biographies, and illustrations throughout these volumes. These are designed to help the reader understand key concepts, and thereby provide a springboard for the next generation of human imagination and sustainability.

References

Anderson, William. 2001. *Economics, Equity, Environment.* Washington, DC: Environmental Law Institute.

Capra, Fritjof. 1996. *The Web of Life: A New Scientific Understanding of Living Systems.* New York: Anchor Books.

Collin, Robert William. 2007. *Battleground: Environment.* Westport, CT: Greenwood Press.

Collin, Robin Morris and Robert William. "Where Did All The Blue Skies Go? Sustainability and Equity: The New Paradigm." *Journal of Environmental Law and Litigation* 9 (1994):399–460.

Dubash, Novraz K., and Daniel Bouille. 2002. *Power Politics: Equity and Environment in Electricity Reform.* Washington, DC: World Resources Institute.

Johnson, Steven M. 2004. *Economics, Equity and the Environment.* Washington, DC: Environmental Law Institute.

Paehlke, Robert. 2008. *Democracy's Dilemma: Environment, Social Equity, and the Global Economy.* Cambridge, MA: MIT Press.

ACKNOWLEDGMENTS

We would like to express our gratitude for all those who helped us with this encyclopedia. Willamette University's President Lee Pelton, Professor Joe Bowersox, and the Center for Sustainable Communities provided foundational support. We are also grateful to David Paige at ABC–CLIO Press, for his support, patience, and timely assistance. We are grateful to the students who worked as research assistants for us in this process: Sikina Hasham, and Lacey Lucas. Candace Bolen provided invaluable office support for us for which we are grateful. We would also like to thank all our students over the years. More than a decade of teaching the first sustainability course in a U.S. law school, several sustainability courses in university environmental programs in the United States and abroad, and environmental justice issues to communities and government agencies has exposed us to a wonderful cohort of earnest, thoughtful, and hopeful students. These students are from Auckland University, New Zealand; Cambridge University, UK; Urban and Regional Planning, Jackson State University; Department of Urban and Environmental Planning, University of Virginia; Environmental Studies and Law school at the University of Oregon; Willamette, Tulane, and Lewis and Clarke Law Schools, Hunter College, and Cleveland State University School of Social Work. Many communities and indigenous peoples have also shared their visions of sustainability with us. We are very grateful for their assistance. They include the Choctaw Band, the Spokane Tribe, and the Indigenous Environmental Network. Many community groups and their leaders have shared their hopes for justice and sustainability. They include the Oregon Environmental Justice Taskforce, the National Environmental Justice Advisory Committee, and many community leaders. Federal and state environmental agencies have also made invaluable contributions to this work. They include the U.S. Environmental Protection Agency, the Oregon Department of Environmental Quality, and the Washington Department of Ecology.

All of our students provide us with a window to the future, a future they want to be sustainable. We thank them.

INTRODUCTION

Sustainability is an important concept now developing into public concern and policy all over the world. It challenges us in many ways. It challenges us to account for all environmental impacts, past, present, and future. It also challenges us to examine closely how we treat each other, as well as how we treat the environment. Ultimately, sustainability is an unfolding dynamic of human processes and ecologically sensitive products. This challenges our patience and our understanding. This encyclopedia is written to increase understanding of the concept of sustainability in all its unfolding directions and paths.

The environmental and ecological foundations of most approaches to sustainability and processes of sustainability are its strongest to date. The key principles of sustainability as enunciated by Agenda 21 are:

- Integrated decision making

- Polluter pays principle

- Sustainable consumption and population levels

- Precautionary principle

- Intergenerational equity

- Public participation

- Common but differentiated responsibilities

Each one of these principles of sustainability is undergirded by knowledge of the environment and ecosystems, although some more than others. The soaring interest in sustainability is as much our expanding knowledge of the ecology as it is our cumulative and increasing larger impacts on the environment. It is also one of the most uncompromising foundations of sustainability, giving sustainability proponents and policies impetus to challenge traditions, to change ways of thinking about environmental issues (paradigm change), and to engage in adversarial and controversial issues.

The ecological roots of sustainability run deep. Most cultures were aware that their long-term survival was reliant on natural systems. Only recently have all cultures been made aware that there are limitations on natural systems with the advent of global warming and climate change. More and more people across various stakeholder categories of industry, government, community, environmental, religious, other nongovernmental organizations and labor are able to observe human impacts on the environment. Knowledge of the effects of these impacts on humans and on the ecosystem is slowly emerging. Observations around impacts on humans do not mean "causality," and it is here that many legal and policy controversies develop.

In this volume, we focus on the environmental and ecological roots and basis for sustainability. Issues of business and economics and equity and fairness are covered in the other two volumes. Concepts of what constitutes an "environment" span many cultures, and many cultures suffer from environmental ethnocentrism. Knowledge about the environment is not evenly distributed, and some regions may discover today what others have known for a long time.

The globalization of environmental observation, information, and technological advancement has helped overcome this dearth and disparity of environmental observations. With these increased observations, scientists and others were able to see some of the interconnectedness between land, air, and water natural systems and to relate climate to landscape. Units of complete life systems were called biomes or ecosystems. The globalization of ecological information has been a major push in the social acceptance of sustainability worldwide. Many nations are already seeking ways to reduce carbon dioxide emissions drastically, sequester carbon dioxide, and to accurately evaluate ecosystem carrying capacity.

It is likely that the environmental and ecological basis of sustainability will continue to grow. As governments and communities struggle to make environmental science the basis and parameter for policies around sustainability, knowledge about human impacts on natural systems continues to grow. Knowledge about how to live sustainably will also grow. That is the purpose of this volume.

CHAPTER 1

Overview

This overview presents a quick explanation of sustainability dynamics as they relate to the environmental and ecological basis of sustainability. This explanation is necessary to understand the next section on definitions and contexts for sustainability under environmental and ecological approaches. These concepts are part of the emerging language of sustainability.

WEBS OF LIFE: THE INTERRELATEDNESS OF NATURAL SYSTEMS

All life on Earth is supported by its biological, chemical, and physical processes. These processes configure the elements of nature into systems that support all living things. Over time, humans have developed many different ways to name, identify, and classify these processes and systems.

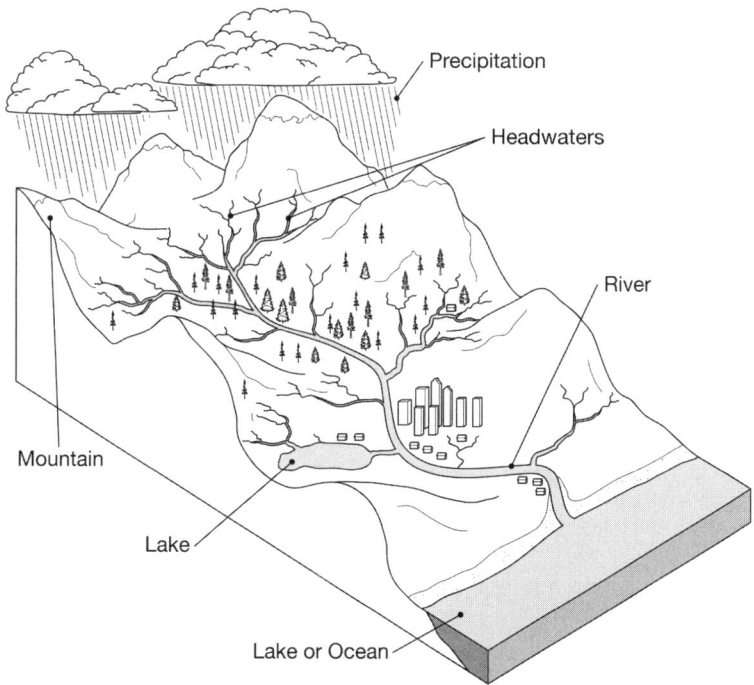

FIGURE 1.3 • Water collects in a basin by draining downhill from various sources. The land area from which water collects is sometimes called a watershed. Outside of North America, watersheds are used to describe the dividing lines between these basins. A water cycle is the interaction of land, air, and water in a watershed. Illustrator: Jeff Dixon.

Different cultures categorize them in different ways. A relatively recent way to think about these webs of life is in terms of ecosystems. Ecosystems refer to constellations of shared conditions that support life in a local area. A particular area on Earth may share water resources, biological species, climatic conditions, and cultural traditions that interact to support distinctively adapted local life. The elements of an ecosystem are dynamic processes themselves, and they interact with each other. For example, within a watershed, processes of evaporation, transpiration, and condensation link land, air, and water.

All human development comes from these basic ecosystems and their services, and all growth is supplied by them. Human development has successfully interacted with ecological functions to meet human needs. We have done this through agriculture by plowing and tilling soil and redistributing water for the growth of plants for food, fiber, and fodder. Hunter-gatherer societies take other species as prey to meet their need for food and fiber. Human uses of ecological functions have grown in scale in response to population growth and because of technology. Human beings have thrived and increased in numbers and complexity of social arrangement because of our successful perturbation and

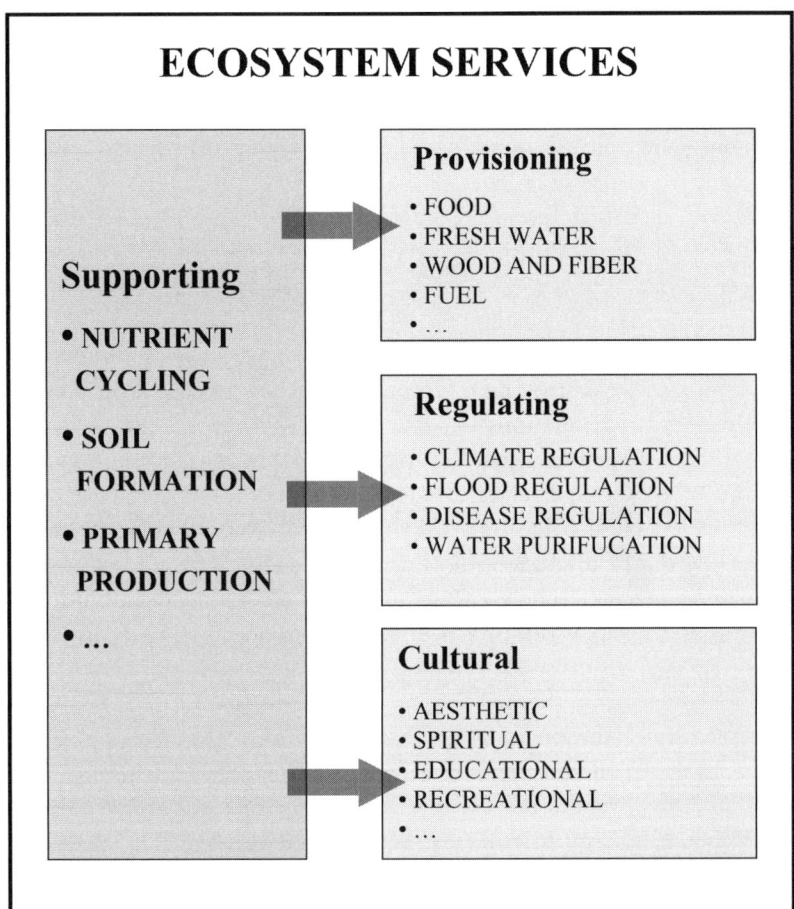

FIGURE 1.4 • Ecosystem services are the benefits that people obtain from ecosystems. Millennium Ecosystem Assessment— www.maweb.org

exploitation of these natural systems. The ability of these systems to produce a flow of goods for human use, however, does not tell us about the health of these systems.

Concerns about the health of ecosystems and the life processes that support them have led scientists to study the baseline conditions of Earth's life support systems. From 2001 to 2005, 1,360 scientists from around the world worked together to assess the world's ecosystems and the services they provide to human development. Their report, called the Millennium Ecosystem Assessment (MA), identified 10 categories: marine, coastal, inland water, forest, dry land, island, mountain, polar, cultivated, and urban. These are not ecosystems in and of themselves; they are each composed of ecosystems. Each of these MA categories shares similar biological, climatic, and social factors that make them distinctive. The ecosystems within each category offer benefits to humans living there. These are referred to in the MA as ecosystem services. Ecosystem services are the benefits people obtain from ecosystems. These services were divided into four broad groups: provisioning resources, regulating services, cultural services, and supporting services.

Human Habitation and Its Impact on the Environment

Human impacts on the environment often come from human uses of the land, air, and water. Waste and pollution are by-products of human development. The life systems of nature produce no products that are not used by other ecosystem processes. Ecosystem processes are thus described as closed loops. As human development increases in scale, so do waste and pollution and their consequences for ecosystems on which all life depends. The challenge of sustainable human development is whether it is possible to develop and grow within the limits of our ecosystems. Is it possible to meet current needs by using ecosystem services without destroying the potential of these systems to supply the needs of future generations?

Human settlements first began to grow near confluences of fertile soil and fresh, navigable water. Water serves to foster agriculture and is a transportation route linking communities together. As communities grow so do discharges of waster and run-off surface water from developed surfaces. These carry concentrations of chemicals, bacteria, and other human waste products into surrounding ecosystems where they can and do cause unwelcome consequences for life systems and living beings within them. The tendency of human settlements to grow causes water pollution of surface and ground water, loss of soil, and salinization of soil; threatens surrounding farmland; and disrupts species habitat. This dynamic is accelerated by the use of fossil fuels to provide fundamental energy to run machinery of every type. Population and technological growth caused major ecological imbalances with many human habitations, later called cities, towns, and villages.

WHY ARE CITIES IMPORTANT FOR SUSTAINABILITY?

Urban areas now occupy about 2 percent of the surface of Earth. The world is rapidly urbanizing, with urban centers become megalopolises. Today there are 19 megacities on Earth with more than 10 million residents. By 2015, some predict that there will be 27 megacities, 18 of which will be in Asia. The environmental impact, or ecological footprint as it is sometimes called, of urban areas is very large. Cities and their commercial, industrial, and residential uses consume about 75 percent of Earth's natural resources. Cities take natural resources from surrounding ecosystem and biomes. They also produce large environmental impacts that can affect ecosystems outside their region. This is one reason the international sustainability community focuses on the growth and development of sustainable cities. *See also* **Volume 3, Chapter 2: Urban Land.**

ECOLOGICAL FOOTPRINT

The ecological footprint analysis of the organization Redefining Progress measures the amount of renewable and nonrenewable ecologically productive land area required to support the resource demands and absorb the wastes of a given population or specific activities. Cities are considered to have large ecological footprints. Density decreases the environmental footprint of human habitations because people live more densely in cities, these areas may be more sustainable than suburban or exurban areas largely because the energy impacts of transportation on the environment are lower. The advantages of density may be lost if land is not preserved from further development. Further, exurban and suburban construction may offset any benefits of environmentally friendly construction on increased commuter impacts leading some to question the conditions for calling suburban or rural home "green." If the energy impacts on the environment, however, are larger because of industrial pollution or because of the costs of moving food and other commodities into the city, then the sustainability of cities is lower. Suburban and exurban areas are spread out and require more roads and have greater impacts on the environment because of vehicle emissions. The sustainability of cities is the subject of much national and international focus and it changes rapidly as knowledge about true environmental impacts of all past and present environmental impacts becomes known, and as changes in technology and human population and consumption patterns change (www.redefiningprogress.org/footprint).

CITIES AND BIOLOGICAL DIVERSITY

Urban areas are also reliant on biodiversity. Biological diversity means the variability among living organisms from all sources and the ecological networks of which they are a part. Biological diversity also refers to diversity within species, between species, and of ecosystems. Biodiversity supports many basic ecosystem functions necessary for life on Earth. Cities rely on biodiversity and healthy ecosystems because they offer resilience to environmental changes. As biodiversity decreases under environmental degradation, cities become more vulnerable to changes because there are fewer reliable regional ecosystems.

Reference

Pelling, Mark. 2003. *The Vulnerability of Cities: Natural Disasters and Social Resilience.* London, UK: Earthscan.

THE CASE OF COASTAL CITIES: WHERE MOST OF THE POPULATION LIVES

Coastal cities with weak ecosystems may be vulnerable to overdependence on regional fishing stocks. The world fishing industry is heavily subsidized by many nations. Their environmental impacts are large. They contribute to damage to seabeds, coral reefs, overfishing, and very inefficient and damaging environmental practices. They can have waste incinerators on their vessels and emit waste into the air as well as directly into the ocean. Ports have traditionally been ecological devastation zones. Ships empty their bilges, spill wastes, and spill products over in ports. Rivers empty their contents into ports and often contain pesticides and fertilizers that have run off from farmlands. Cities that overrely on fishing regional marine ecosystems and accepting high levels of pollution may be compromised in their pursuit of environmentally based sustainability.

Cities are reliant on regional watersheds. Watersheds can range in size from a few acres to millions of hectares. Watersheds cross many lines of political jurisdictions. The provision of water often determines where the official city ends and the squatting community begins. Fresh water is a valuable resource in most of the world. For example, in Nairobi, Kenya, the water comes from the Aberdares Forest. The forest shelters rivers and water catchment areas from the sun, and the roots retain water so that it purifies water and retains water over time. Logging for charcoal threatened to destroy Aberdares Forest, and concern about the long-term sustainability of the logging has delayed it for now.

It is not possible to be sustainable by ignoring cities or any place of human habitation. Some environmentalists' perspectives are antiurban. Right now, more than half the world's populations of more than 6 billion people live in cities. It is predicted that by 2030, there will be 2 billion new city residents. In 2030, 60 percent of the Earth's population will live in urban areas. Also in 2030, it is predicted that there will be 500 cities with a population of 1 million or more. Sustainability advocates in urban areas seek to incorporate environmentally sensitive policy and behavior within the confines of the city. Some have brought urban agriculture into their approaches to sustainable solutions.

TECHNOLOGY AND INDUSTRIALIZATION

Technological innovation has accelerated human development and growth. Technologically driven impacts on our planet's ecological systems are both intentional and unconscious. The internal combustion engine made it possible to farm vast tracts of land and connect communities at great distances. The result is more food for more communities. It also releases greenhouse gases that cause climate change and discharges materials that injure human health, and its basic production and transportation cause contamination of land and oceans. The greatest human impact on ecological functions in both scale and quality has come from industrialization.

WHAT IS INDUSTRIALIZATION?

Industrialization is one type of social arrangement organizing the use of technology. The industrial model of production rapidly consumes resources and energy to produce consumer goods. In industrialized societies, this model has framed society's view of ecosystems as commodities and resources rather than life support systems. Industrialization transforms production of goods from a household-by-household basis into a factory model of specialization. Factory-type production often separates the product from the web of life and ecosystems necessary to create it, again creating a view of consumer products as items of profit rather than products of natural systems. Farming is still a family-operated model of production in many places, with cultivation and animal husbandry embedded in the social arrangements of family life and community cultural structures. When the scale of cultivation and husbandry is embedded in a factory, it becomes agribusiness.

Governments of the industrial age embraced the technologies and social arrangements necessary to create an industrialized economy. The policies of that age facilitated the construction of necessary infrastructure for transportation and labor to produce goods on a scale never before possible. Dramatic consequences came from these policies, especially in the generation of wealth for industrialized countries that often procured the resources necessary for advancement—natural resources and labor resources—by conquest. Along with tremendous wealth, other consequences of these policies include improvements in public health practices, medical advancement, longer life spans, increased population, widespread pollution, and poverty, especially in communities and among people victimized by conquests.

HUMAN IMPACTS ON WEBS OF LIFE

Evidence of the toll human activity has taken on nature's life systems is sobering. Human growth and development have had a significant environmental footprint, and the projected direction of growth raises concerns about whether future generations will be able to meet their fundamental needs from the life systems of our planet. The MA describes the consequences of ecosystem change for human well-being.

It describes the condition and trends in the world's ecosystems and the services they provide. Their four main conclusions are:

1. Human demands have caused a huge loss of life and diversity on Earth.

2. Some human communities have benefited tremendously by the changes made to ecosystems; others have experienced increased poverty that will affect future generations.

3. The degradation of ecosystems and the increase of poverty challenge our abilities to meet goals for eliminating poverty and restoring our environment.

4. Reversing the degradation of ecosystems while meeting increasing demands for their services will require substantial changes.

These conclusions are in the context of human pollution increases in places where global warming will have the most impact on climate changes. These are places between the northern and southern latitudes. Global climate changes in these areas include sea level increases in low-lying countries. Many of these countries are experiencing large population increases and grow limited food supplies along coastal areas. If the oceans do rise, they will destroy areas of food production as the human populations increase. These populations will need to use other land for food production, increasing human impacts on the web of life. In places where wood is used for fuel, these populations will have to scan decreasing forest resources. The downward spiral of poverty and environmental degradation applies here because each dynamic fuels the other.

In other parts of the planet, global climate change can lead to increased desertification. Forests decrease and the lack of tree canopies and root retention of water decrease the ability of the watershed to hold water. In areas where humans live in poverty, this type of environmental degradation can lead to further impoverishment. Poverty makes it difficult to develop adequate resources to prevent ecological degradation. This is one argument that sustainability advocates use when pressing for international invention in poor nations suffering the downward spiral of ecological degradation.

All ecosystems have important biological, hydrological, and physical processes that occur at given times and in predictable ways. They are not yet completely understood in every area. These processes include climatic conditions, soil creation, water aquifers and their recharge rates, and plant and animal cycles. These are important pieces of the sustainability puzzle. Tying this information to human conditions in the area is what is required to understand how humans interact with the web of life, the ecosystem. Human conditions are not always known, and they operate on political and economic cycles. They are often less predictable than ecosystems because of events like war, disease, or migration. If the two systems are not tied together in meaningful ways, the downward spiral of poverty and environmental degradation occurs to the point of irreparable damage to systems of nature on which all future life depends. More concretely, this can mean that the soil erodes or suffers from salinization, lakes lose their fish, and overlogging forever reduces fuel supply.

Given the risks to sustainable development, many nations suffering from the downward spiral of poverty and environmental degradation need both international intervention and greatly improved effectiveness in policy development and implementation. It could require large-scale changes in human behavior, even if that behavior is rooted in longstanding traditions. Because ecosystems are complex, integrated policies toward the environment are needed as opposed to sector-based programs of air, land, and water. This is a challenge for all nations. The size and scale of the policy and its implementation need to match that of the particular ecosystem. These policies also need to include human social policies such as job creation. It is important for sustainable development that all transactions implementing the policy be transparent and completely accounted for. Biodiversity and conservation are often a primary goal in these areas but come at a cost to traditional hunting and wood-gathering traditions. The distribution of benefits and burdens under the emerging policies of sustainable development has to be known and explicitly addressed. Environmental injustices make policy implementation difficult to enforce. Secret water deals, political manipulation of natural resource use, and even conflicts need to be ecologically accounted for. Last, these decisions will be based on much uncertainty. There is so much to know and it is important to begin immediately to stop the downward spiral of poverty and environmental degradation. These new policies of sustainable development must begin without complete certainty and evolve as knowledge is gained through the implementation of sustainable development policies.

Although the nations facing the most dramatic impacts of climate change, population growth, and poverty face the downward spiral of environmental degradation and poverty first, other nations will face similar choices. They will also need to change consumption patterns, develop ecologically based knowledge and policies, achieve transparency in environmental transactions, and move forward under great uncertainty. This does not mean that growth will not occur but that the human impacts on the web of life will not destroy the systems of nature on which future life irreparably depends. ***See also* Appendix D: The Millennium Ecosystem Assessment.**

SUSTAINABLE DEVELOPMENT: GROWTH WITHIN THE WEBS OF LIFE

Sustainable development refers to growth in the human quality of life while not irreparably impairing systems of the environment that future life will depend on. In most Western nations growth often refers to economic growth. This means that the population's per capita and disposable incomes increase. It is usually limited to economic growth for the present generation and does not fully consider economic growth in the future. It therefore generally fails to consider the effects of rapid population growth and the context of future goods and services. Traditional economic growth also fails to consider the systems of nature that constrain economic growth as populations increase and consume more natural resources. These natural resources can run out, become depleted, and fail to replace themselves. Systems of life on which future generations depend become irreversibly impaired and therefore traditional patterns of economic growth are not sustainable.

This is especially true if both population and consumption of natural resources occur simultaneously. A downward spiral of environmental degradation and lowering of economic standards can begin. If consumption becomes a measure of social status, the impact on the environment can be quickly seen. If the economic processes are inefficient and create waste that erodes environmental integrity, then quality of life begins to erode. Waste of all types increases with both population and economic inefficiency. Waste from one source can be a good or commodity to another source. In terms of economic production, reusing what is now waste to develop another commodity is called closing the loops of production to eliminate waste through reuse and increasing economic efficiency. *Waste* itself is a term of art that is carefully analyzed in the process of sustainable development. Wastes can accumulate and overload ecosystems in ways that overwhelm it so much that it threatens to irreparably destroy the systems of nature on which future life depends. For example, it takes tens of thousands of years for nuclear wastes from energy production to safely reenter the environment without causing irreparable damage. Ultimately, ecosystem integrity requires a balance of human needs and impacts with systems of nature over the time scales of nature. As certain ecosystem services like fresh water become depleted, controversies can explode around religion, capitalism, private property, endangered species, and the distribution of environmental benefits and burdens (see Volume 3, Equity and Fairness).

These controversies are described and explored in the context of sustainable development in the sections that follow. Our human understanding of the ecosystem is limited by many factors. There are now only 5,468 mammalian species left and they are the ones that either successfully escape humans or work well with humans. Extinction, a sign of irreparable damage to systems of nature on which future life depends, is occurring at a rapid rate in plant and animal species. Some areas of great biodiversity, such as tropical rainforests, may be losing species before we know they exist. In some ways, our understanding of nature is limited by our own humanity. The human lifespan has increased significantly in the last few hundred years. A few centuries ago humans lived to about 40 years. Now they live to about 75 years.

Scientific frontiers on aging offer the promise of living to 150 years. In terms of lifecycles of ecologic systems, this is a small amount of time. Geologic time moves in millions of years. Species evolution can take thousands of years. Climate change, hydrological cycles, and the complex changes in the web of life on Earth can take hundreds of years. The impact of humans on the ecosystem is now so great that these changes occur within our lifetimes and become more observable. Humans also have the unique ability to accumulate knowledge. We learn from all, or most, of those before us. We can transfer information about the environment from centuries ago. Our geologists can understand climatic conditions that existed millions of years ago. We know that an ecosystem exists, that the web of life is complex, that we have a strong and increasing impact on it, and that we can destroy the very systems of life on which we depend. Sustainable development seeks to preserve the environment so that future generations have the same quality of life, if not greater, than we do.

The United Nations and its various agencies began studying our destructive impact on the Earth in the early 1990s. The concept of global warming alarmed some scientists, and was repudiated by others. Many governments that focused on economic development, free markets, capitalism, and private property also hotly contested the concept. The elected leadership of these nations refused to develop any policy that would decrease the transactional efficiency and corporate profits of the status quo at the time. Powerful nations, such as the United States, rejected international agreements based on global warming concepts. They rejected alternative energy development and engaged in wars to ensure an adequate supply of energy from nonrenewable energy sources such as petrochemicals. Because of the limited scale of human scientific study of the planet, it was hard to tell cause and effect under the limited conditions of scientific inquiry of the time. As global warming became observable and as the effect of humans on global warming became more accepted, the next stage of the debate was on whether global warming caused changes in climate. Even at this time, many powerful nations such as the United States refuse to develop public policies that would mitigate their impacts on global warming. Some scientists pursued the idea that global warming could be economically beneficial for some nations. For example, the melting ice of the Nordic nations might create more hydroelectric power. The primary concern is on exploring economic advantages of climate change, not on mitigating the impacts on ecosystems. This underscores the political context of many scientific inquiries and scientific models of causality. At the turn of the century, the United Nations began the Millennium Project in earnest to bring together the leading scientists of the time to study ecosystems around the globe, especially those that were most at risk of irreparable damage. These scientists embrace uncertainty, transparent environmental transactions, and full cost accounting of all human impacts on the environment. They seek to understand fully the systems of nature on which all future life depends and to develop models of sustainable development. Their work is often controversial because it directly challenges long held traditions around religion, systems of government, and beliefs about the environment.

In Chapter 2 we present a section on definitions of some of the emerging terms. In the area of sustainable development and ecosystem exploration, the current language often falls short of an adequate explanation. Differences in culture, education, values, and experiences with nature all create difficulty in communicating some of these concepts. This is one of the difficulties with defining sustainability. It can be overused in present economic values to market an item without the related concepts of true measure of environmental impacts through the lifecycle of the product. For example, the University of Oregon markets its football games as "sustainable" without a full cost accounting of the emissions from all the transportation, the impacts on the environment from the production of football gear, the impacts on nearby wetlands from all the paving necessary for parking for the stadium, the impacts of the construction of the stadium, and so forth. The abuse of the term *sustainability* does detract slightly from its meaning, but the early definitions still apply and cast manipulated definitions in the light of so-called greenwashing.

Gro Harlem Bundtland

Gro Harlem was born in Oslo, Norway, in 1939. She received her medical degree from the University of Oslo and earned a master's degree in public health from Harvard University. Although her focus was on women and children's health, she became interested and involved in issues of poverty, population growth, food security, and public health on a global scale. She was a well-known doctor of rehabilitation and a prominent member of the Norwegian Labour Party. In 1974, after working for the Department of Hygiene and Oslo Municipal Board of Health, she accepted a position as minister of the environment for Norway and in 1983 was the first woman elected prime minister of Norway. While serving in this position, the United Nations secretary general asked her to create and head the World Commission on Environment and Development. From this post, she wrote "Our Common Future," also known as the Brundtland Report, which is often cited as the foundational document for sustainable development.

The report emphasized several key concepts, including the interconnectivity of the world and the need for increased international multilateralism and cooperation to solve root problems. The report integrated environment and development, "[T]he environment is where we live, and development is what we do in attempting to improve our lot within that abode."

Reference

World Commission on Environment and Development. 1987. *Our Common Future: Report of the World Commission on Environment and Development.* New York: Oxford University Press (available on line at www.un-documents.net/wced-ocf.htm).

CHAPTER 2

Definitions and Contexts

In the context of the environment, sustainability involves many significant terms of art and science to describe complex systems and interactions. The profusion and diversity of these terms can make understanding sustainability in this context confusing to a nonspecialist, even a well-educated reader. Terms discussed in this chapter are placed in clusters relating to land, air, and water, and their dynamic interactions.

Although these terms are defined separately, their contexts are interconnected. Just as the land, air, and water connect aspects of the environment in an ecological context, these terms relate to aspects of the same dynamics of sustainability. The separateness of these terms is a human construct of language but may not adequately reflect the interconnectedness of nature. One of the main challenges to defining any terms of sustainability is to describe environmental dynamics for which there are no words. All human cultures have words that describe dynamics unique to that culture, whether it is the many types of snow described by Eskimos or types of sand in sandstorms described by the nomads of the Sahara Desert. Humans define and refine terms of the environment based on their collective experience with it. Because sustainability collects all human experience with nature, there are not yet words that transcend all human experience to bring meaning to the terms that currently surround the word sustainability. These terms come from environmental or ecologically based sources of sustainability. They are often based in the sciences—biology, chemistry, mathematics, physics, and so forth. Other approaches to ecosystems and to environments can be more spiritually based, but refer to the same dynamics as Western science.

Understanding environmental dynamics requires many observations of natural systems over time. These observations can come from scientific, indigenous, government, and community sources. As environmental information develops and ecological understanding deepens, sustainability proponents can bridge cultural divides through a shared understanding of terms.

As the environmental and ecological knowledge base for sustainability expands, language is strained to develop words for new concepts. Imagine trying to describe germs before germ theory was accepted. The idea that something so small it could not be seen by the naked eye could cause scores of deaths was unimaginable. Germ theory also had large economic consequences. If a ship laden with meat and other food had germs, it was quarantined in harbor. People lost money. For many years after germs were discovered, the medical establishment refused to accept it as a valid theory. Medical science served the economic status quo, and the lack of adequate language to convey the environmental dynamics of germs and their spread prevented the advancement of knowledge. This is the current case with concepts of sustainability, sustainable development, and sustainable communities.

This problem with language is why the Millennium Ecosystem Assessment project is so important. The assessment was begun in 2000 by the United Nations and is discussed extensively elsewhere in this volume and in the other volumes. It has been bridging differing scales and epistemologies by linking local knowledge and global science in multiscale assessments since early 2006. Before then, from 2001 to 2005, the Millennium Assessment gathered and applied the work of more than 1,360 leading scientists and environmental experts to evaluate the conditions and trends in the known ecosystems of the world. This preliminary evaluation was the basis for their ideas about ecosystem services, environmental conservation, and sustainable development. They published their findings in five scientific volumes and six synthesis reports. The Millennium Assessment has helped create a context for action around the word "sustainability," even though there is still much uncertainty around the use of the term.

Ecological interactions, chemical synergies, and feelings of both personal responsibility and hopelessness dominate current areas of uncertainty. Climate changes may develop "environmental refugees" as climate change increases desertification along equatorial food sources. Many readers interested in sustainability begin with an interest in its environmental and ecological foundations. One obstacle to language development in areas of sustainability is epistemological uncertainty. Many people do not even know the questions to ask. What environmental risks are cumulative? How do you actually do an ecosystem risk assessment? How will fast climate changes affect me?

The term *environment* itself is often limited by the political economy of knowledge. Environment can mean many different things to different people. The cultural separation of people from their environment often creates a definition of environment that is separate from their own existence. In some of these instances, the environment is often considered part of a political agenda termed *environmentalism*. In these cases, so-called liberal people promote protection of the environment as part of a social change agenda that threatens economic development.

In other cultures, people are considered an inseparable part of the environment. Some indigenous cultures view their existence as part of an undivided circle of life that includes all past, present, and future lives that are not separate from all nature. In these instances the environment is not viewed as a political stance. It is more related to Western concepts of ecology and spirituality.

These differences in the meaning of the term *environment* become reflected in the social institutions of that culture. Religion, family, work, and education are social institutions that incorporate their meaning of environment. These meanings can become so pervasive they remain unquestioned. They can also affect generational perspectives on the meaning of the term through social institutions that advance knowledge. Higher education is one social institution designed specifically to advance knowledge. Environmental education in Western societies was initially viewed as "liberal." *Environmental sciences* was the first accepted term, with *environmental policy* and *environmental studies* recently coming into greater acceptance. Some researchers challenge whether the environment can fit into current disciplinary structures because the concept of environment transcends disciplines. Because the concept of environment does not neatly fit into current disciplinary structures like history or biology, many scholars in the area of environment do not achieve tenure. The political economy of current disciplinary structures does not allow it because the expansive and undefined nature of the term *environment* does not fit neatly into their particular discipline. Even in Western cultures there are differences in how the concept of environment is awkwardly fit into current disciplinary approaches. New Zealand required a course in the environment and placed it in the philosophy departments of its then 12 universities. Australia also required a course in the environment and placed it in its civil engineering programs in all 32 of it universities. The relationship of higher education (known as tertiary education in these countries) to their environment is reflected in their choice of placement of the environment course. New Zealand's environment is very gentle toward humans; it has no poisonous animals and poses little danger. Australia's environment is very harsh for humans; it having many poisonous and aggressive animals and posing much danger.

Reference

Leal Filho, Walter. 2006. *Sustainability in the Australasian University Context.* New York: Peter Lang.

THE LANGUAGE OF SUSTAINABILITY

The use of the term *sustainable* as encompassing a way of adapting human life to ecosystem limits is relatively recent. It is connected to the environmental consciousness movement of the 1960s and to a series of United Nations conferences and reports beginning in 1972 in

Stockholm Sweden, and culminating with the Brundtland Report in 1987. The idea has been broadly adapted to many uses that reflect competing and sometimes conflicting value choices. These multiple idiomatic uses of the language of sustainability make the idea seem ambiguous. To some extent, that may be an important political value of the language of sustainability: its appeal across difficult value conflicts. But it may also be the weakness of the many usages of this terminology.

In an effort to reveal the basis for different usages of the term *sustainability* or *sustainable development*, we have identified significant contexts for the use of this language in the three main domains that are joined by this term: environment and ecology, business and economics, and equity and fairness. In this encyclopedia, each domain is represented by a single volume. The context for definitions provides a broad picture of how the term is used in each area. By placing these usages within a specific context, one can better understand what value choices are being posed and how they are resolved.

Also, because each domain itself has evolved a specialized and exclusive set of references to sustainability, terms and their usages are rapidly becoming less accessible to a nonspecialized public audience who are nevertheless interested in knowing about them.

A lack of environmental literacy dominates highly consumptive societies like the United States; however, this condition is rapidly changing. One great benefit of science is its ability to transcend cultural differences in areas of environmental observation. One benefit of the globalization of environmental science information is the ability to observe the interrelatedness of global systems. This interrelatedness of natural systems is called "the web of life."

ATMOSPHERE

Earth's atmosphere is approximately seven miles of air surrounding the planet and extending into outer space. In the atmosphere water and air interact with each other and with the land below.

The atmosphere can hold a certain amount of water in its gaseous form; the amount varies with temperature. When air becomes saturated, often as a result of temperature changes, condensing water can fall from the sky to Earth in many forms of precipitation.

The atmosphere is a dynamic air mass, integrally related to land and water bodies. Environmental degradation of the air can result in environmental degradation of land- and marine-based ecosystems. Environmental degradation of the atmosphere is also becoming increasingly observable with the increase of satellite monitoring. Natural events such as volcanic eruptions can load the atmosphere with so much particulate matter they darken large parts of Earth. Human impacts on the atmosphere, such as air pollution, ozone depletion, and overdevelopment, can also load the air with particulate matter. Because the atmosphere is so

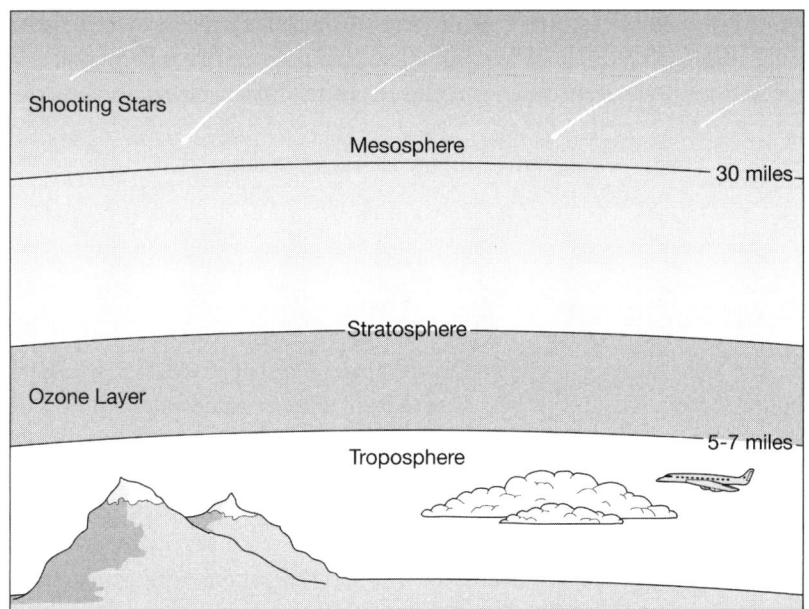

Figure 1.5 • Layers of Earth's atmosphere. Illustrator: Jeff Dixon.

integrally related to land and water systems, this particulate (and other chemicals) matter eventually affects land and water also.

The ecosystem impacts of atmospheric degradation greatly concern sustainability advocates. Environmental monitoring of the atmosphere is now able to locate many more major sources of air pollution.

Reference

Gunn, Angus M. 2003. *Unnatural Disasters: Case Studies of Human Induced Environmental Catastrophes.* Westport, CT: Greenwood Press.

Atmospheric Deposition

The flow of water moving from the sky toward land carries dissolved material within it. When that flow results in dissolved material being deposited on the Earth's surface, whether on land or on the sea, the process is called atmospheric deposition. When the dissolved materials are pollutants or toxins, their effect on ecosystems can be large. Pollutants or toxins can be atmospherically deposed from the air to water through wet or aqueous deposition, dry deposition, or gas absorption.

Many toxins are persistent in the environment. That means they can last a long time before they are broken down into their least harmful components. When these pollutants are atmospherically deposed, they can spread over large areas and affect large parts of an ecosystem. If these chemicals hit large bodies of water and the air is warm enough to cause the water to evaporate, these pollutants and toxins can revolatize back into the air. This can occur over and over again until the air is not warm enough to cause evaporation. The most toxic chemicals are called persistent bioaccumulative toxic substances (PBTs). They can also accumulate in wildlife, causing reproductive problems and other harmful effects. With global warming there is concern that increased air temperatures spread the PBTs to distant places such as the Arctic.

Reference

Smil, Vaclav. 2008. *The Earth's Biosphere: Evolution, Dynamics, and Change.* Cambridge, MA: MIT Press.

Climate Change or Global Warming

As the balance of gases in the atmosphere tips, the Earth's atmosphere holds more heat, and as the temperature of our atmosphere rises, climate zones around the world are changing. The causes of climate change include human introduction of greenhouse gases into the atmosphere. How much of climate change is human driven and the speed of climate change are controversial subjects. Virtually all scientists agree, however, that climate change is occurring, and a significant percentage of that change is human driven, or anthropogenic.

Climate changes are of extreme importance to sustainability advocates because they provide evidence of the limitations of Earth's natural systems. Climate change affects water, ecosystems, food, coasts especially, and the health of land animals, especially humans. The Intergovernmental Panel on Climate Change summarized their predictions of the impacts on these systems by degree of global mean annual temperature change in the following figure.

Global warming is predicted to increase water availability in moist tropics and higher latitudes, decrease water availability and increase drought in mid latitudes and semiarid low attitudes, and will expose hundreds of millions of people to water stress. Other predictions include increasing species extinction with increase in temperature, increasing coral bleaching and mortality, and increasing wildfire risks. In terms of food sheds, scientists predict that global warming will cause tendencies for cereal productivity to decrease in low latitudes and tendencies for some increase in cereal productivity in mid to high latitudes unless it gets warmer faster than currently predicted, in which case, productivity will decrease in all latitudes. Rising ocean levels from global warming will be a significant part of climate change. Increased damage from floods and storms is predicted, resulting in a loss of as much as 30 percent of the wetlands. In terms of health, predictions are for an increased burden from malnutrition, diarrheal, cardiorespiratory, and infectious diseases; increased morbidity and mortality from heat waves, floods, and droughts; and changes in the distribution of some disease vectors.

As evidence mounts of the limitations of Earth's natural systems sustainability, proponents hope this awareness translates into meaningful approaches to public policy and private behavior changes toward sustainability.

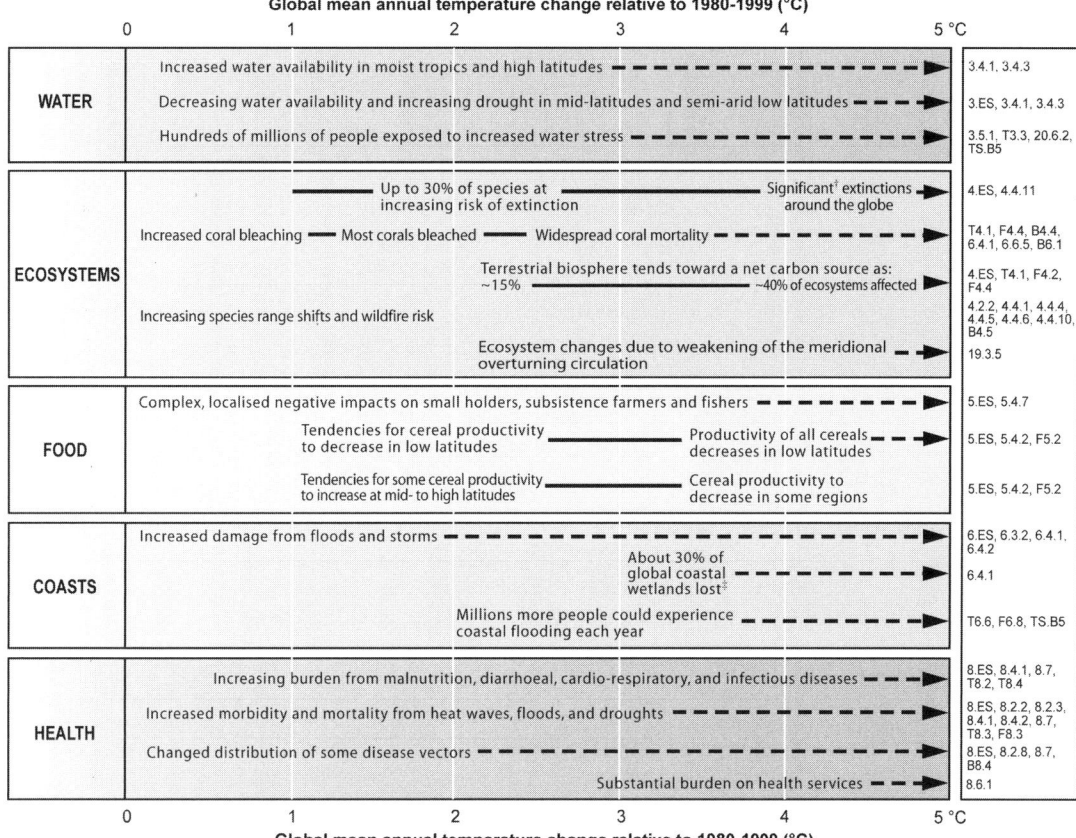

Global mean annual temperature change relative to 1980-1999 (°C)

FIGURE 1.6 • Key impacts on natural systems as a function of increasing global average temperature change. Reprinted from Intergovernmental Panel on Climate Change, Climate Change 2007, Impacts, Adaptation and Vulnerability: Working Group II Contribution to the Fourth Assessment Report of the IPCC (Geneva, Switzerland: Intergovernmental Panel on Climate Change, 2008), 10.

Reference

Parry, M. L., O. F. Canziani, et al., eds. 2007. *Climate Change 2007: Impacts, Adaptation and Vulnerability. Contribution of Working Group II to the Fourth Assessment Report of the Intergovernmental Panel on Climate Change.* Cambridge, UK: Cambridge University Press.

Evaporation

When water changes form from liquid to gas, the change is called evaporation. The rate of evaporation is part of most models of climate change. The rate of water evaporation is also an issue in irrigation of agriculture. Traditional canal systems can lose water through evaporation. When water is tightly allocated, the rate of evaporation can determine whether a landowner gets water that year because there may not be enough to supply everyone.

Fossil Fuels

Humans burn materials formed during prehistoric periods as sources of energy for most of their basic needs. These fossilized materials are tens of thousands of years old and must be taken from the crust of the Earth. They are rich in hydrocarbons, a type of complex molecule formed of hydrogen and carbon. When these materials are burned, they give off energy in the form of heat, gases, and particle matter that enter the atmosphere. These gases include greenhouse gases that can cause Earth's atmosphere to become warmer. These particles include volatile organic compounds that can cause danger to human health.

Petrochemicals are now used in the production of many goods. Many plastics, cosmetics, and building materials have petrochemical components. Petrochemical dependence is a motivating force for sustainability because many nations strive to be energy independent. Reliance on another country for a nonrenewable resource, like gas and oil, often stifles the economic development and political independency of that nation. The search for alternative energy sources could result in more sustainable approaches. *See also* **Volume 1, Chapter 4: Climate Change**.

Greenhouse Gases

When fossil fuels are burned, they produce energy and gases. Some of the gases released into the atmosphere cause chemical and thermal reactions. Nitrous oxide is a greenhouse gas that has been linked to the acidification of rain and oceans. Carbon dioxide is another greenhouse gas linked to climate change.

Most of the increase in global average temperatures since the mid 20th century is likely due to increases in the emissions of gases caused by human activities. Global greenhouse gas emissions have grown 70 percent from 1970 to 2004. Carbon dioxide emissions have grown about 80 percent during the same period.

One of the biggest concerns many sustainability advocates have regarding the atmosphere is the effect of carbon dioxide on the erosion of the ozone layer. The ozone layer protects the Earth from harmful ultraviolet radiation from the sun. A major part of the ecological footprint of cities and organizations is their carbon emissions.

Many organizations and meetings now make it possible to buy carbon offsets. Carbon offsets are a relatively new concept. They were developed by private organizations as a way to mitigate carbon impacts of a given activity. For example, conferences that involve a large number of airplane trips have large carbon impacts. Sponsoring organizations sometimes offer carbon offsets for an additional fee. Carbon offsets have grown from conference options to one of the considerations for environmentally conscious financial investing. Local, state, and federal government levels in the Untied States are just

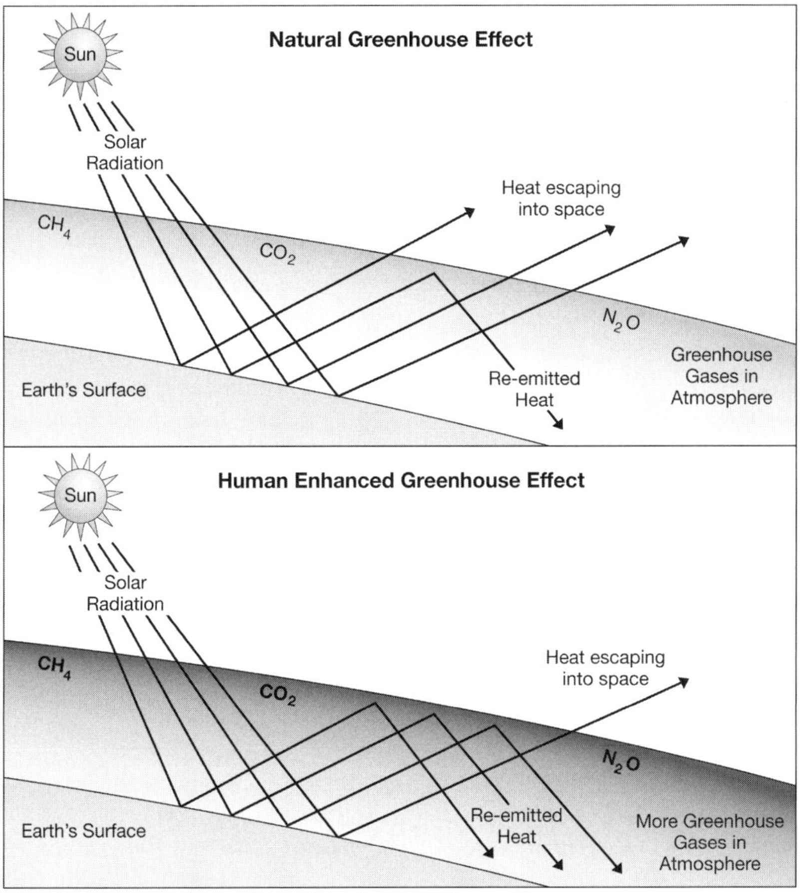

FIGURE 1.7 • How green-house gases trap heat in the Earth's atmosphere. Illustrator: Jeff Dixon.

beginning to consider carbon offsets as part of environmental public policy.

Carbon offsets can be somewhat controversial because it is questionable whether they do in fact offset the environmental impacts of carbon emissions. Others have criticized them because their volume is too small to make a difference if done on a voluntary level. There are other proposals of carbon dioxide sequestration that would theoretically pump carbon dioxide emissions into underground storage areas. These areas would have to be very large, and the amount of energy to move that amount of carbon dioxide might itself have negative environmental impacts. Some research indicates that it may be possible to sequester large amounts of carbon dioxide in a type of rock called peridotite. This type of rock removes carbon dioxide from the air and stores it as an inert chemical as limestone and other carbonates. The process would work by pumping carbon dioxide down bore holes hundreds of feet deep into peridotite. The peridotite would be exploded to create more surface area for the carbon dioxide, and heat would be added to begin the chemical action of making the carbon dioxide inert. Scientists speculate that once started, the fire would continue on its own. Some areas of the world have large areas of peridotite. The nation of Oman has approximately 2,400 square miles that could sequester about 4 billion tons of carbon dioxide a year. *See also* **Volume 1, Chapter 4: Climate Change.**

Ozone

Ozone is a type of oxygen molecule that is distributed throughout the atmosphere of Earth. In the levels closest to the land, such as the troposphere, ozone molecules harm the breathing of animals. At higher levels of the atmosphere, such as the stratosphere and above, ozone reflects dangerous types of radiation from the sun back into space away from the earth. The amount of ozone in the near levels of the atmosphere has increased because of human activities. This anthropogenic source of ozone comes from consumer products and industrial uses that use ozone as a propellant from deodorant sprays to rocket fuels. These uses cause ozone to accumulate in the troposphere and deplete ozone in the stratosphere.

Ozone-depleting chemicals of halons, chlorofluorocarbons (CFCs), hydrochlorofluorocarbons (HCFCs), methyl chloroform, carbon tetrachloride, and methyl bromide were internationally regulated by the Montreal Protocol. Since the early 1990s, these ozone-depleting gases have decreased about 20 percent.

References

DiMento, Joseph F. C., and Pamela Doughman, eds. 2007. *Climate Change: What It Means for Us, Our Children, and Our Grandchildren.* Cambridge, MA: MIT Press.

Emanuel, Kerry. 2007. *What We Know about Climate Change.* Cambridge, MA: MIT Press.

Volk, Tyler. 2008. *CO₂ Rising: The World's Greatest Environmental Challenge.* Cambridge, MA: MIT Press.

Transpiration

Water can move from its liquid form to a gas by means of thermal and biological processes. This flow is called transpiration. When that process occurs with trees, the air can get filtered, carbon dioxide can be consumed, oxygen can be produced, and the temperature can decrease. The effect of global warming on climate change and the effect of tropical deforestation on that process are controversial. Deforestation for logging, to clear land for grazing and for development of human habitation, is occurring at an unsustainable rate.

Volatile Organic Compounds

Some chemical compounds can be vaporized and can enter the atmosphere in the form of gases. This process can take place at room temperature or in the range of temperatures of many industrial processes. As these chemicals volatize they can present a threat to workers, community residents, and the environment depending on the toxicity of the chemical and the exposure to it.

Volatile organic compounds represent one of the early, big steps in environmental policy development around air quality. A range of volatizing chemicals have been introduced, and there is an environmental and industry debate on almost every one. Environmentalists and most sustainability advocates claim more chemicals should be regulated.

BIODIVERSITY

There are three types of biological diversity. The variety of plants, microorganisms, and animal life on Earth is the subject of ongoing study. We do not know how many types of life Earth hosts, even while we are losing some species every day as a result of planetary changes. This is one type of biological diversity. Another type of biological diversity is the range of ecosystems. A third type of biological diversity relates to the genetic diversity within a particular species.

The loss of biological diversity relates to loses in all three types of biological diversity. They are a great concern for advocates of sustainability with an ecological focus because they are occurring at many times the natural rate. According to one United Nations report, more than 3,000 wild species show a consistent decline of 40 percent from 1970 to 2000. Inland, generally freshwater species have declined in population by about 50 percent. Other species have declined about 30 percent The same report concluded that between 12 and 52 percent of present day bird species could be extinct. Because not all bird species are documented, it is impossible to say when exactly these species could be extinct. This highlights one of the problems of habitat loss for sustainable ecosystems. Some species may become extinct before they are known, or before they can be protected. The main reason for extinction is habitat loss, especially the loss of tropical rainforests. Invasive species also cause loss of biodiversity by displacing native plants. Climate change may also increase loss of biological diversity through species loss. Some species may not be able to adapt to rapid changes in climate. The environmental impacts on the largest habitat on Earth, the deep ocean, are still unknown.

This loss of biological diversity has large implications for ecologically based sustainability advocates. In some ways, it motivates individuals and governments to take stock of their environment so they can consider how to act sustainably. Some scientists project that the loss of genetic and species biodiversity is already well on its way. The main characteristic of surviving species is how well they get along with humans. An unknown and controversial aspect to preserving all types of biological diversity is the use of technology. Another site-specific question is how much biodiversity is necessary to support the local ecosystem or biome? How much carrying capacity does the local ecosystem have?

A fundamental policy question for ecologically based sustainability proponents is how much biological diversity is necessary and what kind of biological diversity do they want to sustain.

Reference

O'Riordan, Timothy, and Stoll-Kleemann, Susanne, eds. 2002. *Biodiversity, Sustainability and Human Communities: Protecting Beyond the Protected.* Cambridge, UK: Cambridge University Press.

United Nations Secretariat of the Convention on Biodiversity. 2006. Global Biodiversity Outlook. Montreal: Secretariat of the Convention on Biodiversity.

Wildlife Conservation Society. 2009. *State of the Wild: A Global Portrait 2010–2011.* Washington, D.C: Island Press.

Jane Goodall

Cognitive ethology is the study of the thought processes of nonhuman minds. Jane Goodall is a British primatologist who pioneered the study of chimpanzees in Tanzania in the 1960s. She did intensive field studies of the tool-using and tool-making abilities of chimpanzees that refuted long held views that humans were the only tool-making species. Her first field studies were accomplished as a young woman when many thought it was too dangerous for a young woman to go into the African wilderness.

Dr. Goodall earned her Ph.D. in ethology from Cambridge University in 1965 after her first set of field studies in Tanzania. After that she returned to Tanzania and began the Gombe Stream Research Center. In 1977, she founded the Jane Goodall Institute for Wildlife Research, Education and Conservation. Dr. Goodall has won many international awards, honorary degrees, and honors. She has also published many books and articles. She has taught at many universities including the University of Southern California, Stanford, and Cornell University.

Figure 1.8 • Dr. Jane Goodall of England, world-acclaimed conservation biologist and pioneer for her work on chimpanzees in Africa, delivers her keynote speech while a picture of a chimpanzee, which she calls her oldest chimp friend, is projected in the background, Tuesday, December 7, 2004, in Singapore during the Biology in Asia international conference. AP Photo/Wong Maye-e.

Ex Situ and In Situ Conservation

Ex situ preservation of species refers to preserving plant and animal species outside their naturally occurring habitat. There is a long history of collection and conservation of species by both private individuals and large institutions. These collections were often amassed without concern or knowledge of the ecologies or cultures from which they came. Preservation of these species in their natural habitat is referred to as in situ conservation. In situ conservation requires engagement with issues of ecology and culture in the place of origin. *See also* **Volume 1, Chapter 4: Ex Situ Conservation.**

Fisheries and Fishing

An early warning of how large human impacts can be relates to the depletion of marine fisheries. Humans have always fished. Fishing is a way of life, recreation, and spiritual need. Technology has so changed human proficiency at fishing that Jacques Cousteau, the legendary ocean explorer, described it as a transition from hunting to harvesting. Large ocean trawlers with nets drifting for miles and catching everything in their path can quickly deplete fishing reserves. Good fishers once had to know where the fish were, but radar and other fish-finding devices have made this knowledge no longer necessary.

Fish population counts, enforcement of fishing regulations domestically and internationally, and strongly held fishing traditions all combined to create an area of controversy. The marine fishery populations, however, have declined by such large numbers that the evidence of severely negative environmental impacts from overfishing, undereinforcement of domestic and international fishing regulations, and cultural rigidness is indisputable. Fishing populations have all generally declined. Large predator fish such as tuna have declined by 90 percent of their original fishing stocks. Human impacts are decimating many other marine environments. Coral reefs and large groves of mangroves are negatively impacted by excessive fertilizer and silt runoff. About 20 percent of coral reefs globally are gone, and mangroves have suffered large and immeasurable loss of habitat.

Species of fish are found in the environment in specific localized areas, unless they are farmed or stocked by humans. The stock of such locally found species of fish are called fisheries. Humans hunt these stocks for food and sometimes for fuel and fiber. Fisheries are sometimes described as a type of commons.

New theories about fishery practices challenge traditional assumptions about stable fish populations. The challenged assumption is that most fish populations are naturally stable, and that overfishing can be counteracted by throwing back small or pregnant fish. This was challenged by research on California sardines applying complexity theory. By leaving behind juvenile fish and catching the big fish, populations became even more unstable. Juvenile fish did not have enough fat to withstand changes in their environments and were subject to large population losses. Fish populations are naturally unstable, but this challenges traditional fisheries management theory based on the maximum yield that implies the fish harvest can be adjusted at an equilibrium rate that keeps yields stable. *See also* **Volume 1, Chapter 4: Tragedy of the Commons; Volume 2, Chapter 5: Shares.**

References

Culver, Keith, and David Castle, eds. 2008. *Aquaculture, Innovation, and Social Transformation.* New York: Springer.

Webster, D. G. 2008. *Adaptive Governance: The Dynamics of Atlantic Fisheries Management.* Cambridge, MA: MIT Press.

Genetically Modified Organisms (GMOs)

The study of genes and associated sciences has developed to the point of permitting humans to create new species of plants and animals by combining their DNA in new ways. This kind of manipulation of plant and animal life has raised many ethical questions, popular controversy, and scientific questions about long-term consequences.

Habitat

The kind of plant life, water systems, and other ecological configurations necessary to support a particular plant or animal's life is called its habitat. It is a physical place that is necessary for a plant or animal to live and procreate. It can be more than one place, as in the case of migratory species like some fish and birds. It is the place the plant or animal lives.

The loss of many plant and animal species is often attributed to loss of habitat. When wilderness areas are opened up for oil drilling, as in the case of the Arctic Wildlife Refuge, the habitats of migrating species such as the caribou is damaged. Species that follow the caribou, such as wolves and mosquitoes, are also affected by habitat loss. Approximately 15,000 to 30,000 species extinctions occur because of human activities and their contribution to habitat loss. Aquatic species are also affected by habitat loss. In the United States about 20 percent of freshwater fish species and about 50 percent of mollusks are either extinct or are threatened with extinction because of habitat loss.

Human population growth, with its increased impact on the environment, is a direct cause of habitat loss. Human actions like agribusiness, clear cut logging, massive mining operations, water control systems like dams and diversions, building construction, and road construction all damage habitat. Animal and plant species can no longer find food and water in these former habitats and become threatened with extinction.

Habitat preservation is one of the main reasons for the development of state and national parks. Wilderness areas stress habitat preservation, and their advocates resist all mining, oil drilling, logging, and road development. The protection of endangered species in wilderness areas is controversial because wilderness areas are designed to protect the habitat of threatened species, even unpopular ones that can cause economic loss to nearby property owners. An example of this is the protection of wolves in national parks.

ANIMAL LANGUAGE, COMMUNICATION, AND MUSIC

Most current scientific studies of biodiversity examine population counts. Sometimes they examine migratory paths, their niche in a given ecosystem, and their evolutionary pathways. They seldom examine the way animals learn, communicate, and think. How animals communicate may affect their ability to survive, which affects their population.

Loud noises increase stress levels for most animals including humans. Noise affects how animals react to their environment. Human settlements, with airports, sirens from emergency vehicles, and high background noise, generally drive animals away. Loud noise in nature can mean danger, especially if it is a strange or unusual noise. It can affect how animals breed, migrate, and find food and shelter. If these activities are hindered, the animal populations tends to dwindle. When the numbers of animals decrease, other aspects of the ecology can be affected. Plants dependent on certain animals for procreation, such as those dependent on bees and butterflies for the transfer of pollen, may be negatively affected. The overall biodiversity of a given ecosystem suffers, and if it suffers irreparable damage then chances for sustainability are nil.

Research into the language and music dimensions of animals is just beginning. It is also very controversial, with some linguists maintaining that language is the sole area of humans. The language abilities of animals depend on the animal. Whales and elephants are known to communicate long distances using infrasonic sound imperceptible to humans. Dolphins and birds communicate using a variety of sounds. Research on prairie dogs shows that they communicate using key aspects of language—nouns, verbs, and adjectives. Prairie dogs inhabit the same colonies for many years. They are the favorite prey of almost every hunter in their ecosystem. Prairie dogs communicate about individual coyotes who hunt them. They communicate with each other about whether the human hunters have a gun. They can communicate with each other about the hunting activity of specific hawks.

ANIMAL MUSIC

The differences and similarities between language and music are complex. Music tends to have a relationship between sounds that form certain ratios that form patterns. Music is made up of melody, rhythm, meter, tonality, and timbre as the primary patterns. Some languages such as Mandarin Chinese are tonal in character. Some researchers believe that animal music may predate human music because whales predate humans. Only about 10 percent of all primate species make music. Some whale species use songs that repeat refrains and that rhyme. Bird-songs have intrigued humans for centuries and have distinct meanings. The famous composer Mozart kept a starling as a pet. He altered a part of the Piano Concerto in G Major as his pet starling had revised. The bird "revised" it by changing the sharps to flats. Mozart's composition, "A Musical Joke" was written the way a starling sings music. Birds use music very much like humans beyond simple mimicry. Birds use sonatas, variations in rhythm and pitch, accelerandoes, crescendos, diminuendos, and human sound scales.

In terms of biodiversity, animal communication, whether it be language or music, is necessary to understand them and their relationship to systems of nature on which future life depends.

References

Gray, P. M. et al., "The Music of Nature and the Nature of Music." *Science* 291 (2001): 52–4.

Slobodchikoff, C. N. 2002 . "Cognition and Communication in Prairie Dogs." In *The Cognitive Animal: Empirical and Theoretical Perspectives on Animal Cognition,* ed. Marc Bekoff, Colin Allen, and Gordon M. Burghard, 257–65. Cambridge, MA: MIT Press.

Monoculturing

Some varieties of plants and animals become popular at a particular time because of their appeal to businesses and consumers. Some varieties of plants store well for extended periods, making them better for long-range shipping. Some types of animals adjust well to industrialized forms of animal husbandry and meat production. These preferences may take over the cultivation of these types of foods, resulting in the absence or even loss of less favored varieties. These less favored varieties may have many desirable and locally valuable attributes such as drought resistance and resistance to local pests.

When variety is suppressed, the dominant variety may become vulnerable to local conditions and require substantial chemical intervention in the form of increased water usage, fungicides, insecticides, and other toxic substances. Monoculturing of food supplies raises a host of concerns about the loss of locally viable and useful crops.

To respond to some of these concerns, organizations have started to bank seeds from a wide range of plants. A group of botanical gardens and museums has joined with private businesses to catalog and save as much of the cultivars of our plant population as possible in the Millennium Seed Bank Project. This project, managed by the Royal Botanic Gardens of Kew in Great Britain, is an example of ex situ preservation of species, preserving the species outside its naturally occurring habitat. This type of preservation has had a long, increasingly controversial history as applied to zoo conservation of animal species. *See also* **Volume 1, Chapter 4: Controversies, Bio-diversity and Monoculture.**

Reference

Millennium Seed Bank Project. www.kew.org/msbp/visit/index.htm.

ECOSYSTEMS

Ecosystems are groups of natural processes that are connected through basic chemical, thermal, and biological interactions. In Western science, this term is limited to the physical components of a particular region such as the grouping of plants, animals, and other geographic features in a particular area. In other cultures, such a unit would also include intangible and metaphysical features such as the presences of beings that had lived and died in that place, the people who currently lived there, and ancestors.

The space-time qualities of these natural groupings are also described as biomes, ecotopes, and ecotones. These are the basic building blocks of sustainability from the environmental perspective. They are not neatly ordered, and the terms are used loosely. In policy areas, ecosystems are difficult to administer because they can cross local, national, and international boundaries. Most policy approaches to ecosystems examine the watershed and the flora and fauna, and the marine and terrestrial life forms contained in it. Successive and complete food chains, a range of biodiversity, and interdependence of natural systems indicate a healthy ecosystem. Intermittent and incomplete food chains, degraded biodiversity, and tattered interdependence of natural systems indicate an unhealthy ecosystem.

Ecosystem health can often be a matter of the length of time under consideration. An environmental perspective of sustainability defines the relevant cycle of time as that which is necessary for a healthy ecosystem, depending on climate and landscape of the place. Both industry and environmentalists want to know the carrying capacity of a given ecosystem. How much human impact can a given ecosystem take? Environmentally based sustainability advocates and others are curious about their own impact on their ecosystem.

The U.S. Environmental Protection Agency's (EPA) Council on Environmental Quality defined ecosystem as:

An ecosystem is an interconnected community of living things, including humans, and the physical environment with which they interact. Ecosystem management is an approach to restoring and sustaining health ecosystems and their functions and values. It is based on a collaboratively developed version of desired future ecosystem conditions that integrates ecological, economic, and social factors affecting a management unit defined by ecological, not political, boundaries.

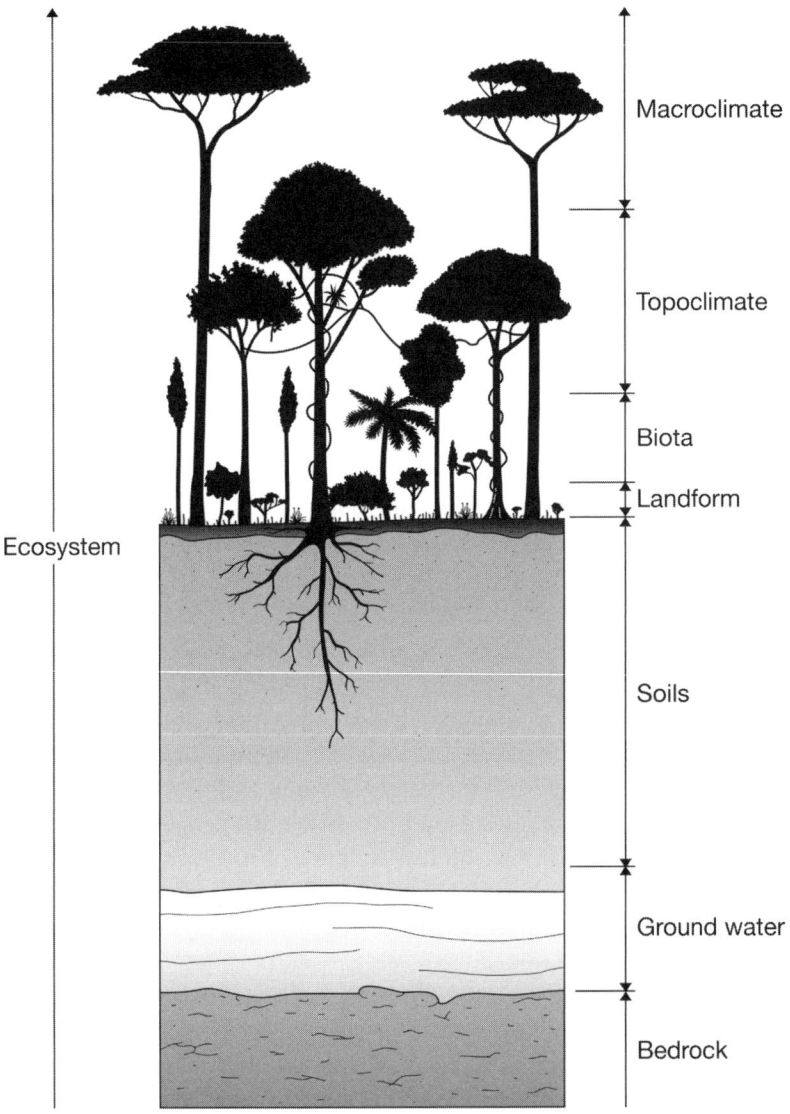

Figure 1.9 • An ecosystem describes the interaction between land, air, water, plants, animals, thermal, biological, and chemical dynamics in a given area. Illustrator: Jeff Dixon.

Macroclimate

Topoclimate

Biota

Landform

Ecosystem

Soils

Ground water

Bedrock

Reference

Gunderson, Lance H., Craig R. Allen, and C. S. Holling, eds. 2009. *Foundations of Ecological Resilience*. Washington, DC: Island Press.

Klijn, Frans. 2007. *Ecosystem Classification for Environmental Management*. New York: Springer.

McLeod, Karen and Heather Leslie. 2009. *Ecosystem-Based Management for the Oceans*. Washington, DC: Island Press.

The Twenty-fourth Annual Report of the Council on Environmental Quality. 1993. Available on line at ceq.hss.doe.gov/nepa/reports.

Carrying Capacity

The idea that the natural processes that support life have limits within which they can operate without failure suggests that excesses in population and/or consumption of resources can deplete our systems. The work of the Millennium Assessment supports this idea. Carrying capacity often tries to allocate population to some resources in order to make our relationship to finite resources more concrete.

The carrying capacity analysis method identifies thresholds (as constraints on development) and provides mechanisms to monitor the incremental use of unused capacity. Carrying capacity in the ecological context is defined as the threshold of stress below which populations and ecosystem functions can be sustained. In the social context, the carrying capacity of a region is measured by the level of services (including ecological services) needed by the community. The strengths of this method are that it is a true measure of cumulative effects against a threshold, it addresses the effects in a system context, and it addresses time factors. Its weaknesses are that it is currently difficult to measure this kind of capacity directly, there may be multiple thresholds, and this type of regional information often is not developed in the United States.

Another method of ecosystem analysis explicitly addresses biodiversity and sustainability. It uses natural boundaries (e.g., watersheds) and applies ecological indicators. Ecosystem analysis entails a broad regional perspective. Its strengths are that it uses regional scale and addresses a large range of ecological interactions (synergy, antagonism, catalysis), addresses time, and seeks sustainability. Its current weaknesses are that it is limited to natural systems, requires more data than we currently have or require, and some of the landscape indicators are still under development. *See also* **Appendix D: The Millennium Ecosystem Assessment.**

Reference

Layzer, Judith A. 2008. *Natural Experiments: Ecosystem-Based Management and the Environment.* Cambridge, MA: MIT Press.

Morrison, Michael L. 2009. *Restoring Wildlife: Ecological Concepts and Practical Applications.* Washington, DC: Island Press.

Ecosystem Services

Ecosystem services come from ecosystem interactions and have uses that benefit humans. These benefits include climate control, fresh water supplies, living soil, plant pollination, basic food production, most raw materials, genetic biodiversity, and medicines. It is difficult to put an economic value on these environmental resources because they are irreplaceable. They are resources that are finite and are used and abused by human impacts. Sustainability proponents want to minimize consumption of essential natural resources by measuring and understanding ecosystems and ecological limitations. This requires natural resource conservation and enhancement, and a strong public policy approach around sustainable resource management of soil, water, air, agriculture, energy, and transportation. By valuing ecosystem services, sustainability advocates hope to push for cleaner production processes for goods and services and to greatly reduce waste.

The excessive use of ecosystem services is most prevalent in the developed nations of the world, where approximately 25 percent of the population consumes roughly 75 percent of the global resources. Ecologically based sustainability will be a big challenge for North Americans and may require major and unforeseen lifestyle changes.

Environment in International Application

The term *environment* can have different meanings when used in an international context. It does not always include ecosystem considerations. It can mean environmental security, which can refer to adequate food and water supplies. It can mean environmental health, which includes human rights. The line between environmental protection and human rights shifts in the international context. Environmental rights can mean adequate medical care for the human population. Forced relocation of people, the effects on vulnerable populations, the effects on women and children, and the rights of indigenous people fall into these categories. Vulnerable populations generally refer to the young, sick, and elderly. From an international transactional perspective on the environment, human rights are considered on a project level, not a national level. Environment can also mean the way environmental impacts are handled in large international financial transactions. In these contexts the ecosystems are given greater consideration.

There is no single, unified international standard of practice or measurement for impacts or for monitoring of these impacts on the environment. There are international treaties and agreements, United Nations Standards, the World Bank's Pollution Prevention and Abatement Handbook, the World Health Organization suggestions, and the standards of individual countries. Some specific agreements that can be used are the Malpol Convention, the World Heritage Convention, the Aarus Convention, the Ramsar Convention, the Washington Treaty, the Red List of the International Union of Concerned Nations, and directives regarding resettlement and indigenous peoples of the World Bank. There are many other international agreements, formal and informal, that circumscribe the term *environment*. As world concern about sustainable development increases, more focus is brought into processes such as the Millennium Project.

Differences in the meaning of the term *environment* create confusion, controversy, and conflict as nations seek to increase the quality of their citizen's lives and other nations seek to decrease environmental impacts that irreparably destroy systems of nature on which future life depends.

All these agreements, treaties, and standards are difficult to enforce and monitor. Many use the term *environment* in ways that conflict with each other. One of the main differences is whether human rights issues, ecosystems assessments, and community involvement are considered. When large projects are developed that require international sources of revenue, however, some of these issues can be seen more clearly.

In large international financial transactions *environment* is a term that is often controlled by the source of the financing. Most large financial lenders now want to prevent any adverse environmental impacts and to control them when they happen. They will either deny the loan, or if negative environmental issues arise, suspend loan payments. Some lenders go further and want to achieve environmental improvements. By environmental improvements they mean environmental conservation, consideration of alternative energy applications, and protection of the global environment. Protection of the global environment generally refers to greenhouse gas emissions, especially carbon dioxide. In terms of the application of the term *environment* to decision making, most large lenders profess to want these processes to be transparent to all stakeholders. This generally means that public notice of the decisions and the ability to comment on the decision are available to the general public in the place of the environmental impacts.

Environmental assessment and monitoring are supported by international financing of big projects. These can be political and controversial. Nongovernmental organizations are now recognized by many international agreements and treaties and will often monitor and assess the environmental impacts of a project. Given the emphasis on transparency, it is now difficult to ignore these groups and they could derail a big project. Most international sources of funding want the project proponent to first review the environmental impacts on the project. They also require the project proponent to monitor the project for environmental impacts. An open question is whether they require the project proponent to mitigate the environmental impacts, and whether those environmental impacts will be monitored. Since the term *environment* can mean impacts on human rights, these dynamics can be very political. Most lenders require that the environmental assessment procedures of the host nation be followed and be made available to the impacted community. Some require consultations with the host community that are part of the project plan.

A big question surrounding the term *environment* in international transactions is who the stakeholders are. Generally they are the local community and local nongovernmental organizations. Sometimes they are all those who are at risk of being adversely affected by the project. Because the term *environment* can include human rights, those adversely affected can include a larger group of people than those who are on exposure vectors of emissions. They can include people in contiguous nations. This also heightens the political context, but may also expand the environmental impact assessment into broader ecosystem considerations. An open question is whether to consult with communities and nongovernmental organizations outside the host country.

In international projects with environmental impacts, there is great concern about the flexibility of the standards. Will they take into account the vulnerability of the natural environment? If the human culture is one of corruption, will they still allow the project to go forward even if community monitoring indicates that it is unsustainable? How available and useful will the monitoring information be? If it is not available or useful, will it stop the project? Does monitoring include studies about ecosystem impacts of potentially endangered species? Would these studies be conducted by an employee or consultant hired by the project proponent or by an independent source? Currently, inmates at the Cedar Creek Corrections Center in Littlerock, Washington are successfully helping to save the endangered Oregon spotted frog (*rana pretiosa*) by raising them behind bars.

These are difficult and unavoidable questions engendered by the ambiguity of the term *environment* on an international level. There is concern from environmentalists that these ambiguities can be used to exploit areas that are not yet aware of their levels of ecosystem capacities. In areas with little or no human habitation of large project development, it is unlikely that the term *environment* is used. Sustainable development, however, will require knowledge of these areas to determine their level of vulnerability and carrying capacity. ***See also* Appendix D: The Millennium Ecosystem Assessment.**

I = PAT

I = PAT is a formula that helps conceptualize the carrying capacity of an ecosystem. It stands for: impact on environmental systems equals the product of population, affluence, and technology.

The more affluent a population, the more resources it tends to consume. Technology affects use of resources as well. Some technologies significantly increase the use of resources. For example, consumption can also be influenced by technology changes like the introduction and proliferation of personal computers and cars. Technological impacts may also be affected by the cultural and social behaviors such as privacy, conspicuous consumption, and individual ownership of personal property rather than group sharing. Some technologies may also arguably decrease the use of resources. The facsimile machine decreased the environmental costs of communication when compared to hard copy document retrieval, delivery, and storage, for example.

Sustainability advocates who support the I = PAT approach often do not challenge growth or capitalism directly. Economic growth, in their view, can be consistent with enough environmental preservation if strong guiding regulations enforced by government act in a decisive manner. They want green growth along with traditional and some innovative environmental policies. The assumption of I = PAT is that technological changes in environmental mediation and remediation will occur so that it compensates for the additional environmental stresses caused by the growth of population and other technologies. A criticism of this approach is that it will require developed nations to reduce environmental impacts faster that economic growth, and faster than the free market can produce the large-scale changes in technology across an entire spectrum of human activities. *See also* **Volume 1, Chapter 4: Population and Consumption.**

HUMAN VALUES AND GOALS

Conservation

This philosophy emphasizes appreciation of nature, avoiding waste, and preserving natural resources for human betterment, including spiritual, physical, and economic improvement. Some conservationists reject the idea of exploitation of undeveloped natural sites for profit. Others believe that commerce and conservation can mutually coexist.

Reference
Thoreau, Henry David. 1854. *Walden: Life in the Woods.* Boston: Tickner and Fields.

Ecology and Deep Ecology

Ecology and deep ecology are philosophies that place humankind on an equal level with other organisms situated in the context of life systems. Humankind and our interests are not viewed apart from, or superior to, those of other beings or other parts of our natural systems. This philosophy leads to an increased focus on interconnectedness of phenomena.

One of the original authors of the term is Arne Naess who used it in his article, "The Shallow and the Deep, Long-Range Ecology Movement: A Summary." Deep ecology considered human-centered approaches to the environment as shallow and arrogant. Shallow ecology views the environment as an instrument to be used for the benefit of humans. The modern environmental movement, with its emphasis on parks and conservation, is considered shallow ecology. Deep ecologists view the environment as both the ends and means of all human involvement. This means that all life has equal value and cannot be differentially valued as better or worse when compared to other life forms. A second principle of deep ecology is that there is no boundary between human self and the environment. Because all life is valued the same, any division between life forms, including humans, is shallow ecology.

In the mid-1980s, deep ecologists developed an eight-point platform to define their goals.

1. The well-being and growth of human and nonhuman life have value in themselves. These values are independent of the usefulness of nonhuman life for humans.

2. Biodiversity contributes to these values and are values in themselves.

3. Humans have no right to reduce biodiversity except to satisfy their own vital needs.

4. The growth of human life and culture is compatible with a much smaller human population. Furthermore, the growth of nonhuman life requires a smaller human population.

5. Current human interference with nonhuman life is excessive and getting worse.

6. All policies must be changed. This must occur in all areas—economics, technology, and ideologies.

7. The dominant ideological change is to increase appreciation of the quality of life as opposed to always increasing the standard of living.

8. Deep ecologists have an obligation to try, directly and indirectly, to implement these necessary changes.

Deep ecology and its principles are often criticized. One of the most frequent criticisms is that it is impractical. It cannot be applied when deciding competing and conflicting environmental disputes. An example of this criticism is that the last of a given species cannot be treated as equal to a species that is dominating. Another criticism of deep ecology is that it ignores the sentience of humans. The ability to think for other species, to consider whole ecosystems, and to evaluate our impact on those ecosystems make us responsible in a way that would not be so if all species, including humans, were absolutely equal.

Aldo Leopold

Aldo Leopold (1887–1948) is considered the father of wildlife ecology. He began his life in Burlington Iowa, where he spent much of his time in nature. In fact, he almost did not graduate from high school because he spent so much time and energy outdoors! In the end, however, he went to Yale to study forestry and ended up teaching game management at the University of Wisconsin. Leopold was concerned about the disappearance of wilderness space in America. In response to his concern, he helped create the first wilderness region regulated by the National Forest System.

His most famous achievement was his book *A Sand County Almanac,* in which he introduced his idea of "land ethic":

[A] land ethic changes the role of Homo Sapiens from conqueror of the land-community to plain member and citizen of it. It implies respect for his fellow-members, and also respect for the community as such.

The land ethic simply enlarges the boundaries of the community to include soils, waters, plants and animals, or collectively: the land.

A land ethic, then, reflects the existence of an ecological conscience, and this in turn reflects a conviction of individual responsibility for the health of the land. Health is the capacity of the land for self-renewal. Conservation is our effort to understand and preserve this capacity.

References

Leopold, Aldo. 1949. *A Sand County Almanac.* New York: Oxford University Press.

Leopold, Aldo, The Wilderness Society, www.wilderness.org/aboutus/leopold.cfm.

Reference

Naess, Arne. "The Shallow and the Deep, Long-Range Ecology Movement: A Summary." In *Inquiry: An Interdisciplinary Journal of Philosophy and the Social Sciences* 16 (1973): 95–100.

Ecopsychology and Biophilia

Ecopsychology means the connection between human psychology and ecology. Human psychology examines behavior, motivation, thought processes, and feelings of humans. Ecology is the relationship of land, air, water, and all the living things within a given geographic area. Ecosystems do not always have solid boundaries, but human political and geographic systems almost always do. Human relationships with ecology differ widely and pose a significant challenge to the development of public policies and private behaviors around sustainability. In 1992, Theodore Roszak wrote, *The Voice of the Earth*, describing the term *ecopsychology*. In that book he describes the alienation of humans from their environment. He argues that we must recognize our implicit and explicit relationship with nature in order for societies to place a higher value on sustainability. Many ecopsychology theorists posit that technology plays a strong part in increasing human alienation from the environment. Technology increases human alienation with the environment by altering the focus of humans to things other than Nature, such as watching television. Technology also increases human alienation with Nature by valuing Nature as a means to an ends that degrades Nature. For example, with heavy construction equipment and advances in water control technology, what were wetlands necessary for wildlife habitat become swamps that must be drained for natural resource extraction and real estate development. The increasing urbanization of human settlement patterns also increases the distance between humans and nature by removing all Nature from where we live, work, learn, play, heal, and worship.

Human survival has often meant buffering and separating human life from the forces and patterns of the environment and ecology of the place inhabited even as we manage them for our benefit. In developed societies, this separation often exists at a high level that is reflected in not knowing or thinking about the source of our food, the cost of our lifestyles, and the amount of waste we generate and where it goes. This separation and disconnection from the natural forces and cycles are also thought by some to be responsible for a human psychic and mental conditions ranging from anxiety and depression to more dangerous and dysfunctional conditions. A branch of psychology and therapy has developed this idea into a practice of treating humans by reconstructing connections between individual humans and the ecology of the places they live. Human alienation, or lack of involvement with the people and places around one, can be considered pathological if the alienation poses a threat to the individual or those around them. In seeking to help or cure these individuals, ecopsychologists try to recreate innate bonds with nature.

Reconnection with the natural forces that govern our well-being on Earth is essential to knowing and safeguarding those forces so that they may continue to benefit lives that depend on them. Destructive and dangerous adaptations occurring from climate change, acidification of the oceans, and loss of species are considered the result of a failure to understand human relationships within and to the ecosystem providers that all life depends on. Ecopsychologists examine ecological sciences to analyze the human mind as an integral part of Nature. In this view sustainability includes an environmentally based version of sanity. Grieving for the loss of Nature, personal and cultural identity through excessive material consumption, human addiction and technology, skills of ecological perception, the deconstruction of whiteness and the end of racism, and species arrogance are all topics covered in the emerging realm of ecopsychology.

Biophilia refers to the innate closeness humans have for nature. The tendency of humans to focus on life is considered a biologically based need. According to this theory, human development and human species development are basic, biological needs. As justification for this theory, researchers point out that fear of poisonous species develops very quickly, almost instinctually, in humans. In contrast, however, fear of more dangerous and life-threatening objects is slow to develop and almost always requires learning and experience to develop into a fear. These objects include guns, knives, and cars. Another set of research that underpins this theory is that humans prefer to look at water, green plants, and flowers rather than the human-built environment of glass, brick, concrete, and steel. Yet another research set argues that the development of language and thought processes relied on the use of natural images found in the environment.

Biophilia as a theory makes some strong claims. First is that it is biologically based, that it is somehow inherent in our physiology and genetic code. This makes it part of humans' evolutionary development. This implies that our development as a species was intrinsically related to the development of the growth of our development with nature. It also implies that human competitive advantages over other species and our genetic strength improved because of our innate relationship with nature. On a psychological level, biophilia implies that our emotional needs and thought process improved. The most intriguing implication of biophilia, and one with great implications for sustainability, is that our environmental and conservation ethic for nature is a product of self-interested genetic evolution. This is thought to be particularly applicable to human concerns for biodiversity.

Both ecopsychology and biophilia underscore the relationship of humans to the environment on psychological and evolutionary levels. To deny that humans are both psychologically and evolutionarily engaged with the environment around them would require any observer to overlook all our history. To prove exactly how humans are both psychologically and evolutionarily engaged with the environment around them, however, has so far escaped the models of causality and proof current scientific approaches demand. These concepts are extremely important for public policies and private behaviors of sustainability. To the extent scientific models of proof and causality prevent the growth of the fields of ecopsychology and biophilia, they also prevent the development of meaningful policies and personal growth in the area of sustainability. Nonetheless, strong social forces are driving sustainability, and among them are the need for people to reduce their alienation from nature and to increase their understanding of their place in nature as a species among many.

References

Nicholsen, Shierry Weber. 2003. *The Love of Nature and the End of the World: The Unspoken Dimension of Environmental Concern.* Cambridge, MA: MIT Press.

Roszak, Theodore. 1992. *The Voice of the Earth.* New York: Simon and Schuster.

Roszak, Theodore, Mary E. Gomes, and Allen D. Kanner, eds. 1995. *Ecopsychology: Restoring the Earth, Healing the Mind.* San Francisco: Sierra Club Books.

Sabatier, Paul, et al., eds. 2005. *Swimming Upstream: Collaborative Approaches to Watershed Management.* Cambridge, MA: MIT Press.

Wilson, Edward O. 1984. *Biophilia.* Cambridge, MA: Harvard University Press.

Environmental Justice

Environmental justice refers to the disproportionate impact of environmental decisions based on race or income. Environmental decisions are often new types of decisions for governments, and they are always political. Those groups in society that are marginalized or disenfranchised from political paths of power will often receive the burden of many environmental decisions. Those groups who control government and have access to power will often avoid the burdens of many environmental decisions while receiving the benefits. The environment does not care about justice or unfairness but reflects the totality of human actions in ecological systems. (These principles are discussed in great detail throughout volume 3 of this encyclopedia, which focuses on equity and fairness.)

Major inequalities in environmental protection and decision making have left their mark on nature. After the Emancipation Proclamation, racial discrimination in voting rights, housing, education, and transportation remained. The legal remedy for ending racial discrimination relies on proof of the intent to discriminate. This is difficult if not impossible for two reasons. First, it is difficult to prove in court what is in a person's thought process leading up to an "intention." It is almost impossible to prove that an organization like a business or a city has the "intent" necessary to validate a charge of racism by U.S. legal standards. Second, the courts themselves can be racist. There can be racist judges, racist juries, racist lawyers, and racist prosecutors. Nature responds only to the results, not human standards of intent. Cities and other communities of color have been environmentally ignored and treated as depositories of wastes for hundreds of years. Simply expanding the current and recent system of environmental protection to treat all communities equally will not be enough to remedy past environmental injustices and, therefore, not sufficient to reach conditions of ecological sustainability. Even if current policy did treat all communities environmentally equally, the failure to remediate the cumulative effects of past environmental degradation will only continue to cause degradation for that bioregion. The uncontested disparities in exposure to environmental hazards by race continue to adversely affect and accumulate in people of color. Nature registers all our interactions with it; humans cannot and sometimes will not do so.

Environmental justice focuses on the benefits and burdens of all past, present, and future environmental decisions. The motto of environmental justice communities is, "We speak for ourselves." Sustainability examines the carrying capacity of the land, air, and water—of the ecosystem. In many cities, it is likely that the carrying capacity of ecosystems is exceeded. The environmental burdens are borne both by the people who live, work, play, learn, and worship there, and borne by nature. In cities spanning different ecosystems, where one has exceeded its capacity and another has not, the question will move to a sharing of benefits. It could be that surrounding suburban neigh-borhoods may have to shoulder some of the environmental burdens of neighboring urban communities that have exceeded their carrying capacities. More studies are investigating the concentration of waste and pollution with associated human health impacts in communities of color and poor communities. When environmentally hazardous activities are shared, the result tends to be a reduction and elimination of waste and pollution in all communities. This is an express goal of sustainability, but one fraught with a history of racism and scientific uncertainty. It will require difficult dialogues. As noted by one researcher:

> The idea of race exists because people give it particular meaning, a meaning that changes with time, place, and circumstances. How-ever, one constant remains—the privileging of whiteness through different devices, social patterns, and even laws. This racial position-ing is maintained in part through an unwritten rule that it cannot be discussed. In fact, the corollary rule mandates that we talk about the social desire for equality while avoiding an examination of white privilege or any other privilege.

Stephanie M. Wildman, 1996 When privileges are unknown, they are unacknowledged by those who hold them; however, the environ-ment reflects these results of unequal, disproportionate decisions. It poses a thorny and controversial area for sustainability advocates.

The international context of environmental justice also focuses on the distribution of benefits and burdens of environmental decisions. This includes not only the results of those decisions, but also the right to participate equally in the decision-making processes. The international context of environmental justice can be extremely adversarial, as in the case of water from Palestine going to Israel. Without water resources in an arid, desert climate, a nation is not likely to prosper. They are therefore unlikely to give up a scarce resource even if it is not theirs. Other countries have systems of government that have exploited classes of their populations and in so doing abused the environmental areas of the exploited groups. India has a caste system with levels of discrimination solidly in place in all policy areas, and in many other institutions such as religion and family. For many years, South Africa had a horrific system of racial oppression called apartheid. Africans and other dark-skinned people were forced to live, work, play, worship, and learn in environmentally degraded places while suffering extreme burdens. The benefits of environmental degradation went to a small group of white people who had colonized the area. In the international context, environmental injustices often form the backdrop for a global push for sustainability. *See also* **Volume 1, Chapter 4: Environmental Reparations.**

References

Committee on Environmental Justice, Institute of Medicine, National Research Council. 1999. *Toward Environmental Justice: Research, Education, and Health Policy Needs.* Washington DC: National Academy Press.

Initial Report of the United States to the United Nations Committee on the Elimination of Racial Discrimination. 2000. Available at www.state.gov/www/global/human_rights/cerd_report.

Wildman, Stephanie M. 1996. *Privilege Revealed: How Invisible Preference Undermines America.* New York: New York University Press, xi.

Environmentalism and Environmental Ethics

Environmentalism adds concerns about species, their habitats, and ecosystem-wide perspectives to the idea of conservation. Environmentalists value natural systems for their intrinsic qualities, not only in terms of their instrumental value to humankind.

THE U.S. ENVIRONMENTAL MOVEMENT

The U.S. environmental movement is a social movement to protect and conserve the environment. It is partially defined by the meaning given to the term *environment,* which is often based on the personal values, cultural background, and life circumstances of an individual. Most people do care about protecting and conserving the environment, but when it comes at a personal cost to them, they may not. One common definition of environmentalists that defines part of the U.S. environmental movement is one who uses a political approach to stopping land development and pollution. The U.S. Council on Environmental Quality defines the human environment as the natural and physical environment and the relationship of people to that environment.

Environmental ethics in the United States emerged as a way to guide human behavior toward the environment.

HISTORICAL OVERVIEW OF U.S. ENVIRONMENTALISM

The history of the U.S. environmental movement has been determined by the awareness of the environment. As Western Europeans moved west and settled in places that were either not inhabited or inhabited by indigenous peoples, their awareness of the environment greatly increased. Before the advent of mass communication, ideas were disseminated by writers and artists who were able to publish and sell their works. Their ideas and thoughts that transmitted a new awareness moved slowly back east to Washington, D.C. These writings and artworks often inspired and thrilled potential settlers to move out West. The view of the environment was that it was an exciting, albeit threatening, place. Nature was there to serve the people brave enough to conquer its elements. It was not until the mid-1880s that the public attitude toward the environment changed significantly. The massive killing of the buffalo almost to extinction made people aware of the fragility of the environment. Buffalo numbered in the millions and were killed for fun and profit. George Perkins Marsh published *Man and Nature* in 1864 and changed the public's awareness that they could irreparably damage the environment. Later that century, public awareness of the environment increased to the point of political actions. Ideas and thoughts were also transmitted much more quickly now, and western states had more influence in national politics. John Muir founded the Sierra Club to conserve nature. The organization helped develop areas to limit human impact, such as Yellowstone and Yosemite parks. These areas were protected for their natural beauty, but they were considered worthless in terms of developmental value. By 1916, the U.S. National Park Service operated 37 parks. Although environmentalists were successful in creating parks, some thought the parks could be used for planned natural resource use and others wanted them protected forever as wilderness areas. The difference in these views has significant implications for modern approaches to sustainability.

Conservation approaches to the environment saw a revival under Gifford Pinchot. Conservation to Pinchot meant management of resources with the goal of returning the most benefit to the people. In 1898, Pinchot was appointed head of the Forestry Division of the U.S. Department of Agriculture. In 1905, all forest preserves were placed under the Department of Agriculture. Pinchot was adamant that forests were for people. As Theodore Roosevelt stated in advocating for the creation of national forest reserves:

> And now, first and foremost, you can never afford to forget for a moment what is the object of our forest policy. That object is not to preserve the forests because they are beautiful, though that is good in itself; nor because they are refuges for the wild creatures of the

wilderness, though that, too, is good in itself; but the primary object of our forest policy, as of the land policy of the United States, is the making of prosperous homes. It is part of the traditional policy of home making in our country. Every other consideration comes as secondary. . . . You yourselves have got to keep this practical object before your minds; to remember that a forest which contributes nothing to the wealth, progress or safety of the country is of no interest to the Government, and should be of little interest to the forester. Your attention must be directed to the preservation of the forests, not as an end in itself, but as the means of preserving and increasing the prosperity of the nation.

Alfred Henry Lewis, 1906

Pinchot's view of the environment allowed large areas to be logged, mined, and grazed. His vision and policies had a large impact on the environment. Nonetheless, Muir-type environmentalists continued in their political efforts to protect the environment from human impact that caused degradation.

Public awareness of the human impact of the environment greatly increased in the 1960s. Rachel Carson, a scientist, wrote *Silent Spring*, a book about the ecological impacts of pesticides. The Cuyahoga River in Cleveland caught fire in 1969. There was a massive oil spill off the coast of Santa Barbara, California, on January 28, 1969. The public was becoming concerned that not only was the environment deteriorating; it was doing so in a way that could affect public health and safety. The rate of technological growth and waste production in the petrochemical and nuclear industries raised the specter of potential global destruction. On January 1, 1970, President Nixon signed the National Environmental Policy Act, which formed the Environmental Protection Agency. This law eventually required environmental impact statements for activities that significantly affected the environment. This raised public awareness even more. In the late 1960s and early 1970s, many political environmental organizations greatly increased their membership base. The Sierra Club formed a legal defense fund. The Natural Resources Defense Council was designed to litigate environmental issues. Many new environmental laws, such as the Clean Water Act and the Clean Air Act, were designed to allow citizens access to federal courts to litigate their claims. These high-profile and often controversial cases created a foundation of new rules and regulations protecting the environment and increasing both controversy and awareness in states and municipalities.

In the 1980s, this increased awareness developed into more radical environmental groups that engaged in protest activities. Groups such as Earthfirst, Greenpeace, and the Sea Shephard Conservation Society engaged in confrontational protests to protect environments. The unequal distribution of environmental benefits and burdens also became known and forged the Environmental Justice movement. Although the choice of George W. Bush in 2000 was a large step backwards in environmental protection, awareness of the impact of human development on the environment became better known all over the world. The environmental priorities of the Obama administration embrace this expanding awareness of environmental impact, based on science and law. Rita Jackson, the nominee for Administrator of the Environmental Protection Agency, underscored this in her first memorandum to EPA employees on January 23, 2009.

This explosion of knowledge about environmental impact directly fuels the overwhelming public support for sustainability. With population growth, knowledge growth, and large-scale environmental impacts observable within one human lifetime, awareness of the degrading environmental impacts of past human practices is stimulating the development of sustainability.

Many in the U.S. environmental movement believe that a radical new approach to environmental sustainability is needed because of the failure of their efforts to protect the environment. Others feel that the environmental movement has increased in power and sophistication, but not enough to prevent environmental degradation. Members of the major U.S. environmental organizations point out that the environment is everyone's responsibility, not just advocacy groups.

The condition of the environment affects environmentalism. Some feel that environmental activism has reached a tipping point, where change must occur or the choices become limited. Many of these choices excluded ecologically based sustainability. The rate of change is what startled many environmental activists and organizations. Given our present consumption patterns and with no growth in populations or economies, the rate of environmental degradation would continue. The economy is not holding still but growing by exponential rates. The world level of economic activity in 1950 is estimated to be $7 trillion and that amount occurs every 9 to 11 years. Environmental conditions are associated with other global issues such as social unrest and political power. These are related to uncountable population increases and climate changes. The international and U.S. environmental movements are empowered and emboldened by these dynamics to pursue ideals of sustainability. They start with ideas about ecology and biomes. Given the range of world cultures and religions, however, the discussion turns to environmental ethics. The values that underscore human culture, history, politics, and economics become stark and apparent with increased knowledge about cumulative and increasing environmental degradation caused by human development.

References

Collin, Robert W. 2006. *The Environmental Protection Agency: Cleaning up America's Act.* Westport, CT: Greenwood Press.

Lewis, Alfred Henry, ed. 1906. *A Compilation of the Messages and Speeches of Theodore Roosevelt 1901–1905.* Washington DC: The Bureau of National Literature and Art, p. 208.

McConnel, Grant. 1972. "The Failures and Success of Organized Conservation," in Nash, Roderick, p. 71. *Environment and Americans: The Problem of Priorities.* Huntington, NY: R. E. Krieger.

ENVIRONMENTAL ETHICS

Environmental ethics are rules that determine good behavior toward the environment. A major question is whether the environment is there for humans. Are humans part of the ecology? Is the environment somehow separate from humanity? Many view the environment as there for humans to have, or at least to be stewards of. A major part of human behavior toward the environment is dictated by religion. In 1967, Lynn White Jr. wrote the "Historical Roots of Our Current Ecological Crisis." In that work he held that the environmental degradation that was beginning to also affect humans was the result of the Judeo-Christian worldview. He maintained that these religions are anthropocentric, or human centered. To be environmentally respectful, an ethic should be nonanthropocentric. This view is controversial and directly confronts many worldviews as expressed by religion. (For an interesting perspective on this issue, see Harper Bibles, 2008.) Religious values are among the most deeply held values of any culture. On a fundamental level, they determine what is right and wrong for an individual, and often what is a good or bad public policy.

Environmental ethics go to the underlying values of sustainable decisions, not their practical methods. They closely examine the values of nature, the values necessary to restore an ecologically sound landscape, the values of sustainable development, and the values of a long-term restoration of ecology. Many times they directly challenge the economic policies of nations and communities as short-sighted and as serving the material needs of present generations.

Environmental ethicists attempt to develop ethical theories about the environment. As they do this they try to decide what parts of the environment are worthy of direct moral standing, indirect moral standing, and no moral standing. Values toward the environment play an important part of this discourse, but political power does not. Values under environmental ethics are divided into intrinsic values and instrumental values. Intrinsic values go beyond their instrumental values. They are a means to an end.

There are several perspectives in environmental ethics. The first one is anthropocentrism, which means that only humans have intrinsic value and direct moral standing. All other parts of the environment have instrumental values. Nonanthropocentrism means that humans and some nonhuman entities have intrinsic value. There are several types of nonanthropocentrism. They vary on the extent to which intrinsic value is extended to nonhuman entities, or parts of the environment. Zoocentrism gives intrinsic value to humans and specific nonhuman animals, although interpretations vary as to which entities they consider

Rachel Louise Carson

Rachel Carson is considered the woman who started the environmental movement in the United States. She was one of the forces behind the creation of the Environmental Protection Agency (EPA) and environmental legislation such as The Clean Air Act and The Clean Water Act.

Born May 27, 1907 and raised on a farm in Springdale, Pennsylvania, Carson began as a journalist for the Department of Fish and Wildlife after becoming the first woman to complete the public service exam and worked her way up to become the chief editor for all department publications. After 15 years of service to the department, she resigned to write full time.

She authored several books that introduced the public to wildlife and ecosystems. These books made Carson famous as a naturalist and science writer. In 1962, she introduced the public to the possible harms of indiscriminate use of pesticides with the publication of *Silent Spring*. This book warned of the harmful effects that pesticides could have on the environment, and ultimately humans. Chemical companies tried to stop publication of the book and launched personal attacks on Carson, her work, and her credibility. But the impact of *Silent Spring* was monumental. In the end, the government banned DDT and other pesticides discussed in the book. Sadly, by the time governmental action took place, Rachel Carson had died of breast cancer on April 14, 1964, in Silver Spring, Maryland.

References

Carson, Rachel. 1962. *Silent Spring*. Boston: Houghton Mifflin.

Silent Spring Introduction by Vice President Al Gore, www.uneco.org/ssalgoreintro.html.

Rachel Carson Institute, Commemorating Rachel Carson, www.chatham.edu/RCI/aboutrc.html.

In Memoriam www.rachelcarson.org.

as having intrinsic value. For example, the loss of biodiversity is unethical to some zoocentricists, as it affects sentient animals, not nonsentient animals. Biocentrism gives intrinsic value and direct moral standing to all living parts of the environment. Nonliving entities and ecological systems either have instrumental value or no moral value under this view. Universal consideration gives intrinsic value and moral worth to everything, alive or not. Ecocentrism gives intrinsic worth to biological wholes like species and ecosystems, and not individuals. This is one of the bases of deep ecology.

Reference

Harper Bibles. 2008. *The Green Bible*. New York: HarperOne.

White Jr. Lynn. "The Historical Roots of Our Current Ecological Crisis." *Science* 155 (1965):1203–7.

Millennium Development Goals

The millennium development goals (MDGs) are eight goals outlined by the United Nations. The benchmarks of the MDGs are rooted in international development and are to be carried out by a number of UN member states, as well as international organizations. The MDGs were officially established at the Millennium Summit in 2000 in New York City. At the time, this was the largest gathering of world leaders (eventually surpassed by the World Summit in 2005). The MDGs wanted to prompt progress and set a goal to achieve all eight of the goals by 2015. Thus far, progress has been exceptional by some nations and completely lacking by others. The Multilateral Debt Relief initiative to effectuate debt cancellation for heavily indebted poor countries was in part to help realize some of the MDGs. The eight goals are further divided into 21 targets within the framework of the MDGs.

1. Goal 1: Eradicate extreme poverty and hunger

2. Goal 2: Achieve universal primary education

3. Goal 3: Promote gender equality and empower women

4. Goal 4: Reduce child mortality

5. Goal 5: Improve maternal health

6. Goal 6: Combat HIV/AIDS, malaria, and other diseases

7. Goal 7: Ensure environmental sustainability

8. Goal 8: Develop a global partnership for development

Vandana Shiva

Vandana Shiva is a highly regarded physicist, philosopher, ecologist, feminist, and author. She was raised in India, near the base of the Himalayas. Her father was a conservationist and her mother, a farmer. After receiving her B.S. in physics, she pursued an M.A. in philosophy at the University of Guelph (Ontario, Canada). In 1979, she completed her Ph.D. in quantum theory physics at the University of Western Ontario.

Shiva wanted to be a scientist, but she started to question the relationship between science, technology, and the environment as it related to food and women in India. In 1982, she started her interdisciplinary research at the Indian Institute of Science and the Indian Institute of Management in Bangalore. It was here that Shiva founded the Research Foundation for Science, Technology and Ecology (RFSTE), which focused on the conservation of biodiversity. Shiva also founded *Navdanya* (Nine Seeds) in 1991, which attempts to protect the diversity of India's native seeds. Both RFSTE

and Navdanya support farmers in rejecting pressures to use genetically modified organisms (GMOs) and other technology that endangers India's native seed and plant biodiversity. Shiva argues that "GMOs, such as herbicide resistant and Bt crops, increase the need for chemicals to combat resulting super weeds and super pests, decrease biodiversity through genetic pollution, and destroy farmers' freedom with patent monopolies and dependency on non-renewable seeds" (Shiva, 2002).

Shiva argues for the rights of farmers to use seed from the previous year's crops to produce more crops. With the introduction of patented seeds that resist pests, however, the owner of the patents disallows seed storing. Shiva has also argued against corporations patenting seeds that farmers in India have been using for thousands of years. In a speech regarding the ability to patent life, Shiva expressed her disagreement with corporate practice to slightly alter a plant's genetic code, and then have the ability to patent it as a new product, which was creating more poverty and hunger in India and other developing nations that are agriculture-based.

Reference

Millennium Declaration. www.un.org/millennium/declaration/ares552e.htm.

Sustainable Development

Sustainable development links the goals of traditional development—economic improvement and social betterment—to environmental protection. The most frequently quoted definition comes from a report for the World Commission on Environment and Development called, "Our Common Future," written in 1987 by Gro Harlem Brundtland, then the prime minister of Norway. She wrote that sustainable development is "development that meets the needs of the present without compromising the ability of future generations to meet their own needs." *See also* **Volume 1, Chapter 4: Growth and Development.**

U.S. National Parks and Forests: A Basis for Sustainability?

The concept behind "parks" is often clouded. They are parts of nature set aside for a variety of uses, usually by some branch of the federal, state, or local government. They can be for the recreational purposes of present-day outdoor enthusiasts; reserves of natural resources like timber, minerals, and grazing land; or land held for future real estate development. They can be part of an ecological system, such as a river or tract of land. Parks often divide ecosystems in ways that are arguably nonsustainable. Parks have been created as a tool to drive out indigenous peoples and other landowners. The designation as a "park" does not always mean that the place is based on sustainable development.

The idea of a park to preserve wilderness comes close to protecting a place in nature that provides a platform for sustainability. On August 25, 1916, President Woodrow Wilson signed the Organic Act that created the National Park Service. The National Park Service was an agency within the Department of the Interior. It was responsible for protecting the 40 national parks and monuments then in existence with the intent of preserving them for future generations. The Organic Act specifically states that the National Park Service must:

> promote and regulate the use of [national parks] by such means and measures as conform to the fundamental purpose of the said parks . . . which purpose is to conserve the scenery and the natural and historic objects and the wild life therein and to provide for the enjoyment of the same in such manner and by such means as will leave them unimpaired for the enjoyment of future generations.

> Volume 16 of the United States Code, section 1.

More national parks were established but not all of them were within the Department of the Interior. Different federal agencies had different parks with different rules. There was no "system" of national parks until

1970 when the General Authorities Act was passed. This act unified the national parks into a system and underscored some aspects of sustainability in the following language:

> though distinct in character, [the parks] are united through their interrelated purposes and resources into one national park system as cumulative expressions of a single national heritage; that, individually and collectively, these areas derive increased national dignity and recognition of their superb environmental quality through their inclusion jointly with each other in one national park system preserved and managed for the benefit and inspiration of all people of the United States.
>
> > Volume 16 of the United States Code, sections 1a–1.

These parks were designed and developed for future generations, but were they developed to be sustainable? For many, the impact of modern recreational uses, as well as commercial uses, is destroying the quality of environmental life that is necessary for future generations to enjoy the parks.

An example of this is Yellowstone National Park. The first of its kind, it was established on March 1, 1872. It has many beautiful and unique features such as the geyser Old Faithful and more than 10,000 hot springs. The park itself is a large caldera for a huge active volcano. The 2.2 million acres of Yellowstone is the foundation of the Greater Yellowstone Ecosystem, which is one of the largest intact temperate zone ecosystems on Earth. Many species of animals exist in a delicate, wild balance.

Outdoor recreationists have enjoyed the splendor of Yellowstone for more than 100 years. In 1992, more than 3.1 million people visited Yellowstone. Hiking, swimming, boating, and cross country skiing have taken many recreational users into a wilderness that was sustained. However, technology and commercialism have pushed the envelope of outdoor recreation into areas that have detrimental environmental impacts. Large recreational vehicles, snowmobiles, all terrain vehicles, and motorized water boats can engage Yellowstone in ways never conceived by its founders. The development of roads into the wilderness is the first impact, followed by the increase of emissions into the air, the noise into the landscape, and the waste left behind. The feeding, mating, and growing patterns of the indigenous species are upset with impacts on ecosystem balance.

Snowmobiles have been particularly controversial. Their impact on the environment is large on almost every level. Loud, dirty, and waste-creating; these machines have significant impacts on the pristine snow environment. It is difficult to enforce environmental rules and regulations deep in the woods and mountains of Yellowstone. By the time the proof of significant damage is decided, it may be too late for the ecosystem to reach its previous balance.

Added to the arguments of snowmobilers, that they have the rights to enjoy nature as they wish, are arguments that technology can make snowmobiles quieter and less air polluting, and that snowmobilers will self-enforce waste practices. So far, technology has made snowmobiles more powerful. They are so powerful now that they challenge mountain slopes to create avalanches. Another argument in favor of snowmobiles is that of economic development. Local snowmobile guides, snowmobile manufacturers and distributors, engine repair shops, restaurants, and hotels all want to make money from the increased use of the national parks for snowmobiling.

As a result, the most arguably sustainable of U.S. national parks is under threat by technological advancement in recreational use and the concessions that follow this increased use. These are hotly contested environmental controversies in US National Parks. To many sustainability advocates the national parks represent islands of hope for future generations to know which efforts of past generations were sustainable.

Reference

Collin, Robert W. 2008. *Battleground: Environment.* Westport, CT: Greenwood Press.

Moran, Emilio F., and Elinor Ostrom. 2005. *Seeing the Forest and the Trees: Human-Environment Interactions in Forest Ecosystems.* Cambridge, MA: MIT Press.

Voluntary Simplicity

Individuals who are interested in reducing the impact of their activities on their environment can take personal action to reduce their environmental footprint. Many individual steps are easily incorporated into daily life activities. The impact of these changes may seem small until they are adopted by a significant number of individuals, even if they are less than a majority. The aggregate effect of individual changes can affect changes at a much larger level. Some authors have argued that a "tipping point" for human behavior change exists at 10 percent of the population. Conversely, individuals confronted with the necessity for wide-scale change may feel that their individual efforts lack significance. This psychological state of mind is called lack of self-efficacy.

Many people motivated to change their individual life styles to reduce their impacts on the ecosystems have formed a movement toward voluntary simplicity. As a contemporary movement connected to sustainability, this movement deemphasizes consumer behaviors that consume energy and create waste. It has many other aspects connected to spiritual practices and economic pragmatism.

This movement is voluntary, not required by law or regulation. There is no enforcement mechanism and no certification organization. Nevertheless, a central feature of this diverse movement is the idea that simple living can be fulfilling, rewarding, even rich without excess, and harm to our environment. *See also* **Volume 1, Chapter 4: Green Consuming.**

LAND, SOIL, AND FORESTS

Land is composed of the soil and its products. Minerals, subsurface waters, and vegetation can be viewed as part of the land. About 25 percent of the earth's surface is land. Direct human habitation is on approximately 1 to 2 percent of the land, although the impact of human habitation is far beyond that. Land can often be reduced to a parcel of land with definite boundaries up, down, and around it. These boundaries can be human-imposed political and real estate boundaries or natural boundaries like watersheds. Human-imposed boundaries can be linked to resource extraction and to ecosystem services like agriculture, mining, or fresh water. These boundaries can also be linked to urbanizing actions like real estate development and the built environment of houses, offices, shopping malls, and factories. Land is also an economic concept that forms the basis of private property. In this context it is a bundle of legal rights, responsibilities, and obligations. It has economic value fundamental to capitalism and to liberty concepts. For many indigenous people, land conveys a meaning through a sense of place that gives them a cultural identity.

Most of the earth is covered by water, mainly in the form of salt water. Land is the dry surface portion of the earth's crust. The area of transition from land to water is constantly changing. (See Wetlands and Estuaries defined later.)

Dry land itself has many levels, including the topsoil that is essential to human agricultural cultivation and grazing, as well as deeper levels that contain minerals, oil, and gas. Water resources are also found within these deeper levels. (See discussion of Groundwater.)

Brownfields

Brownfield land or brownfields are abandoned plots of land that were once used for commercial or industrial facilities. Some brownfield sites are the result of decommissioned military or industrial sites. This past use often resulted in contamination by hazardous waste or pollutants. Brownfields are specific plots of land that may be reusable once the land has been sufficiently cleaned up. Most brownfield sites have been left unused for significant periods; however, as land in certain locations becomes more scarce or expensive, brownfield sites become more valuable. Eventually they will be valuable enough to clean up to safe standards and redevelop. Furthermore, with increased precision and new techniques, the ability to bring brownfield land up to safe standards has become scientifically and economically feasible.

The regulation and cleanup of brownfield land is regulated by the EPA. The EPA works with individual states to provide technical assistance and cleanup of brownfields, as well as to help determine sources of funding to ensure that brownfields are given new life.

Brownfield redevelopment is still not perfect and some projects are abandoned because of rising costs resulting from unknown contaminants that exceed the initial evaluation. Most brownfield cleanup projects are for commercial use; however, there are some projects underway to determine whether brownfields can be used to grow crops. The intent is twofold, first to help with the cleanup process of the soil and second to contribute a more efficient production of biofuels.

Greenfield land is either undeveloped land or currently used for agriculture. Some greenfields are greenbelts that have prohibitions against development. Greenbelts are designed to protect the unique character of undeveloped land within areas of extensive development.

Greyfield land is land that was once thought to be economically profitable but eventually became obsolete and outdated. The term is usually applied to areas that were once considered viable retail and commercial plots of land that suffered from lack of reinvestment. Greyfields are not contaminated; instead, they are usually abandoned because of larger developments nearby. Greyfields may have a dormant value because they are often equipped with an underlying infrastructure, like plumbing and sewage systems, that may be used if the land were redeveloped. Greyfields are those plots of land that have become stalled in their industrial development. If the area were to be rejuvenated, greyfields are ideal places to resume commercial and retail investment because much of the basic groundwork may have already been laid. *See also* **Volume 1, Chapter 4: Urban Sprawl.**

References

Cristensen, Julia. 2008. *Big Box Reuse.* Cambridge, MA: MIT Press.

Dixon, Tom et al., eds. 2007. *Sustainable Brownfields Regeneration: Livable Places from Problem Spaces.* Hoboken, NJ: Wiley-Blackwell.

Drainage

Water is commonly removed from wetlands for human development purposes including housing development and agriculture. Some governments have encouraged drainage of vast areas of soil for these purposes. Wetlands are often drained without knowledge or consideration of their functions in the surrounding ecosystems including cleaning of water polluted with nitrates and other runoff from urban and agricultural sources; protection of land from the destructive forces of floods, tides, and winds; and the protection of a variety of plant and animal species.

Reference

Gonenc, I. Ethem. 2004. *Coastal Lagoons: Ecosystem Processes and Modeling for Sustainable Use and Development.* Boca Raton, FL: CRC Press.

Forests

Forests are composed of trees growing, propagating, and dying on land. Forests are an essential part of ecosystems because they

FIGURE 1.10 • Old growth forests contain old living trees, as well as unique biodiversity supported by their longevity. These forests are threatened by logging in many areas around the world. © 2009 Jupiterimages Corporation.

retain water and soil, provide organic material, and provide wildlife habitat.

There are many types of forests. Temperate and boreal forests have a less than 30 percent canopy cover, and less than 75 percent of that canopy is needleleaf and evergreen. There are deciduous needleleaf forests with a less than 30 percent canopy cover, and about 75 percent of that canopy is both needleleaf and deciduous. There are mixed deciduous and needleleaf forests with a less than 30 percent canopy cover, and that canopy is evenly mixed between deciduous and needleleaf trees. There are deciduous broadleaf forests with a less than 30 percent canopy cover, with about 75 percent of that canopy cover comprised of deciduous broadleaf trees. There are freshwater swamp forests, which can have any mixture of tree types present in a water-saturated soil. There are disturbed natural forests, sparse forests, exotic and native tree species plantations, lowland evergreen broadleaf rain forests, lower and upper montane forests, mangrove forests, and thorn forests. There are still tracts of land with unspecified forests. The type of forest makes a difference because of the type of interaction it has with the ecosystem and the amount of ecosystem services it has the capacity to provide.

Forests are affected by many factors such as climate change and human habitation. Fires, acid rains, threats from nonnative species, threats from nonnative insects and animals, and forest land fragmentation all degrade the forest environment.

References

Bachmann, Peter, Michael Kohl, and Risto Paivinen. 1998. *Assessment of Biodiversity for Improved Forest Planning.* New York: Springer.

Binley, Dan, and Oleg Menyailo. 2004. *Tree Species Effects on Soils: Implications for Global Change.* New York: Springer.

Robinson, John, and Elizabeth Bennett. 1999. *Hunting for Sustainability in Tropical Forests.* New York: Columbia University Press.

Infiltration

The movement of water through the soil is infiltration. The amount of water a given type of soil can hold determines how much will run off and how much can be absorbed. The movement of water through soil is controlled by soil porosity, gravity, and capillary action. Decaying organic matter can hold more water than soils dominated by nonorganic matter.

The amount of water a given land mass can hold is important for many reasons. Agricultural productivity, aquifer recharge, and ecosystem regenerative potential are all dependent on how much water a given land mass holds.

Irrigation

The practice of moving water from its locations in rivers, streams, and lakes to dry areas for human use is ancient. Civilizations from the Mayan to the Minoan and up to the present day have used pipes, aqueducts, and other structures to obtain water from remote locations for use in household, agricultural, and industrial applications. Irrigation can be used to change local ecosystems and convert them to more habitable and cultivatable areas. Irrigation can also be used to create artificial ecosystems, sometimes with harmful consequences to the preexisting habitat and detriment to the original water source. Irrigated land can also become saline and depleted when the practice is continued in certain circumstances over a long period.

Freshwater as a natural system is a high priority for sustainability. More droughts and drought-affected areas are predicted. Fresh water reserves stored in ice and snow reserves will decline, and fresh water from mountain melt water will decline. About one-sixth of the global population currently relies on mountain runoff for freshwater. Some areas will see an increase in water.

Reference

Gleick, Peter H. 2009. *The World's Water 2008–2009: The Biennial Report on Freshwater Resources.* Washington, DC: Island Press.

Salinization

When water or land becomes infused with salt, it becomes saline. The process of water or land becoming saline is called salinization. With rising ocean levels into many populated areas, the amount of salt water intrusion into nearby freshwater sources is expected to increase.

The accumulation of water-soluble salts of sodium, potassium, magnesium, and calcium in the soil can occur because of excessive irrigation and could be accelerated because of global warming. Excessive irrigation with high salt waters with large amounts of fertilizers into land masses with low infiltration can be a cause. Global warming may increase the uptake of salt water in the rain that comes from oceans. Salinization is a concern because the accumulation of these salts decreases the productivity of the soil.

Reference

Pilkey, Orrin H. and Rob Young. 2009. *The Rising Sea.* Washington, DC: Island Press.

Sedimentation

Sedimentation refers to the dynamic of how solid particles suspended in a liquid, generally water, collect together. Sediments are often soil particles in water systems that can clog dams, water treatment plants, and industrial processes of mining. Water runoff, soil erosion, deforestation, atmospheric deposition to water, and dredging of waterways may cause sedimentation. Sedimentation can occur far away from its cause. It can cause ecological and economic damage. Many studies of sedimentation examine changes in river flow, water channel modification, endangered species habitat, wetlands, and cultural resources.

Clear cut logging on steep slopes with no riparian protection can cause erosion of soils that can cause sedimentation. This can also overwhelm water treatment plants and endanger public health with unsafe water.

The mouths of rivers, estuaries, and wetlands are often places where natural sedimentation occurs. When ports are located in these places, the sedimentation can block shipping channels and interfere with recreational users. When these locations are dredged to make way for shipping and recreational use, severe environmental impacts can occur.

Urban Sprawl

Sprawl is a contemporary land use problem linked to development patterns. A problem associated with sprawl is that people move out into natural areas with greater and greater impact on the environment and with greater consumption of nonrenewable resources. Three factors fuel sprawl. The tendency of people to reside away from work requires transportation back and forth to work and home. Roads and motor vehicles greatly impact the environment in nonsustainable ways. Second, methods of commuting are inefficient, with many one-driver car trips. This also increases environmental impacts. Third, privacy and status values are reflected in a house, which tends to result in large lot, low density housing. This has a large environmental impact. Sprawl is considered a threat to farmland because residential land is a higher value than most agricultural land, tending to drive farmers out. Sprawl represents an increase in air emissions, decreased energy efficiency, increased water use and runoff, increased solid waste and recycling requirements, and increased nutrient discharges into the water systems. *See also* **Volume 1, Chapter 4: Urban Sprawl.**

References

Curwell, S. R., Mark Deakin, and Martin Symes. 2005. *Sustainable Urban Development.* London: Taylor and Francis.

Davenport, John and Julia L. 2006. *The Ecology of Transportation: Managing Mobility for the Environment.* New York: Springer.

Devuyst, Dimitri et al. 2001. *How Green Is the City?: Sustainability Assessment and the Management of Urban Environments.* New York: Columbia University Press.

Helming, Katharina, Marta Perez-Soba, and Paul Tabbush. 2008. *Sustainability Impact Assessment of Land Use Changes.* New York: Springer.

Monto, Mani, L. S. Ganesh, and Koshy Varghese. 2005. *Sustainability and Human Settlements: Fundamental Issues, Modeling and Simulations.* Thousand Oaks, CA: Sage Publications.

Pelling, Mark. 2003. *The Vulnerability of Cities: Natural Disasters and Social Resilience.* London: Earthscan.

WASTE, POLLUTION, AND TOXIC SUBSTANCES

Waste is a by-product of human habitation and industrialization. As human populations increase, industrialization also increases and both have dramatically increased in the last century. So too have the environmental impacts on land, air, and water. Many of these impacts occurred and are occurring without knowledge. As knowledge about the environmentally degrading environmental and human impacts develops, laws are passed describing what "pollution" is.

WHAT IS POLLUTION?

Pollution is a term of art describing illegal and regulated environmental impacts. In the United States it is narrowly defined by regulations to generally mean emissions of a certain chemical over an allowed amount. This differs from the general understanding of pollution to be any negative environmental impact. Many industries are permitted to emit a certain amount of a given chemical. (For more information see www.scorecard.org, the U.S. Toxics Release Inventory by zip code.) The limited legal definition of pollution is a controversial issue, with implications for sustainable development. Many environmentalists contend that the amount allowed to be emitted is too high and that the list of chemicals regulated is incomplete. The U.S. environmental enforcement system is largely reliant on self-reported information from industry. Of the more than 80,000 chemicals in commerce in the United States, only about 2 percent are tested thoroughly for public health impacts, and very few for ecosystem impacts.

As knowledge about the locations of wastes, the environmental impacts of waste, and the public health impacts of wastes increase, public policy engaging these important issues of sustainability will increase. In chemicals that bioaccumulate in humans and that persist in the environment, scientific knowledge and cumulative effects have pushed the policy envelope.

Reference

Rosemarin, Arno et al. 2008. *Pathways for Sustainable Sanitation: Achieving Millennium Development Goals.* London: IWA.

Persistent Organic Pollutants (POPs) and Persistent Bioaccumulative Toxins (PBTS)

Certain chemicals predominantly used in pesticides persist in the environment long after their application purposes. They can cause environmental degradation of ecosystems. These can cause cancer and birth defects in humans and other species. Even when some chemicals have not been scientifically proven to cause cancer, they can accumulate in organs and tissue all the way through the food chains in a process called bioaccumulation. Some examples of commonly used POPS are phthalates, polybrominated diphenyl ethers, formaldehyde, atrazine, and polycyclic aromatic hydrocarbons. There has been environmentally driven international action on these issues resulting in international treaties to manage and eliminate them. See Stockholm Treaty discussed in Chapter 2.

Risk Assessment

Risk assessment is a form of analysis of the probability and magnitude of harm from various events and activities. It is widely used to make decisions. Insurance relies on this computation of risk of harm in making decisions about whether to insure, and if so, how much to charge for insurance. Risk assessment is also used in decisions about development projects, and it is used by governments in budgets and planning activities.

Related to the science of risk assessment, risk management determines how to plan for and communicate about risks. Risk perception is a science devoted to examining the qualitative aspects of risk, not simply its quantitative aspects.

Government often requires a risk assessment to be performed in many areas of environmental and developmental activity. These studies are used to determine funding priorities. Risk assessment is often done as a matter of expert assessment. However, risk perception is a developing aspect of this science, and a commonly held perception of environmental and ecological risk can provide an important common platform for environmental action that is missing in current controversies. Government has an important stake in developing such common ground as a basis for legislation and regulation for sustainability, especially in changing assumptions of permissions to pollute toward assumptions of eliminating all forms of waste and pollution. *See also* **Volume 1, Chapter 3: Role of Government in Sustainability; Volume 1, Chapter 4: Risk Assessment; Volume 1, Chapter 5: Risk Analysis.**

Sector-Based Approaches

In this approach to environmental regulation of industry, industry is divided into sectors by the Standard Industrial Classification (SIC) code. The United States and many other developed nations use sector approaches to refine the fit between the environmental permit and the industries emissions. Early environmental regulatory approaches relied on a "one size fits all" approach.

Sector-based approaches seek to help industry comply with environmental laws. They theoretically help government understand the complexities of a type of industry so they can work together to reduce environmental degradation. Some international treaties incorporate the SIC codes in establishing comparable environmental regulatory approaches.

From an environmental and community perspective, sector-based approaches are an improvement in most developed countries because they increase regulatory flexibility and transactional transparency. In some industrial nations like the United States, sector-based approaches can reveal if the reach of environmental regulatory authority is enough to act as a platform for new sustainability policies.

WATER

Water is the basis for life on Earth. Fresh water is especially important for life of many animals and plants. With human population increases, increases in water consumption, and increases in waste production, the supply of fresh water is an ecological crisis of sustainability. When a country has less than 1,000 cubic meters per person per year of water, the health and economic development of that country are threatened. If the water is 500 cubic meters per person per year then the ability of humans to survive is in question. In the late 1990s, about 28 nations had water scarcity issues; by 2025 this number is predicted to increase to about 56 countries. The number of people and ecosystems affected will be much greater because of increases in human population and water use. Oceans are also affected by ever increasing pollution. These practices are sometimes called ocean dumping. An example of an ocean dumping practice is when ships empty their bilges, or waste holds, into the ocean. There are many other types of ocean pollution. Plastic pollution is responsible for the Great Pacific Garbage Patch, a patch of floating plastic about 500 miles west off the coast of California. According to the National Academy of Sciences, about 5 million tons of plastic are dumped into the oceans each year. Two other floating plastic garbage patches have recently been discovered, one off the coast of Chile and another off the western coast of Antarctica.

Increases in water use are caused by increases in deforestation, agriculture, mining, and human settlement development. The ability of the hydrological cycle to recharge itself and hold and create fresh water is lowered by the loss of forests, wetlands, and estuaries. Climate changes caused by global warming may also affect the supply of fresh water by increasing ocean levels, floods, and hurricanes. ***See also*** **Volume 2, Chapter 4: Water.**

Acidification

Saltwater in the ocean is increasingly becoming acidic from the deposition of more carbon dioxide. This is causing problems in our oceans parallel to the problems of climate change in the atmosphere.

When water becomes very low in pH, it becomes acidic. pH is a measure of the amount of hydrogen dissolved in water. Carbon dioxide (CO_2) causes increased acidity of water. CO_2, sulfur dioxide, and methane are by-products of burning fossil fuels. When they are released into the atmosphere in the process of combustion, some of the gas mixes with water in the course of a water cycle falling to Earth through atmospheric deposition. It falls as acid rain. Acid rain, together with runoff water and other processes, contributes to killing of living things in lakes (eutrophication). These same conditions contribute to make the oceans and seas more acidic, with the same potential results.

Acid rain increases with petrochemical energy processes, especially coal burning industrial processes for heat and energy. It differs from region to region and can cross national boundaries. The pollution can travel many miles in the atmospheric air currents before it falls to the land as precipitation. Wherever it falls, it can cause environmental degradation. Control of pollution generally and protection of ecosystems especially concern sustainability advocates.

Acid rain has been a political issue on the national and international level since the early 1960s. Many advances in international environmental policy development were spurred by concerns about the environmentally degrading effects of acid rain. Advances in air pollution enforcement and pollution abatement technology reduced airborne emissions in the United States, thereby decreasing acid rain.

Estuaries

Estuaries are areas where fresh water flows into a body of saltwater. Usually these bodies of water are oceans. Oceans hold about 97 percent of the total water on earth. Only .02 percent of the global water supply is in lakes, channels, and seas. An estuary is formed where ocean tides meet river currents, and its unique levels of salinization cultivate species diversity. Estuaries are estimated to provide needed habitat for about three-fourths of the U.S. commercial fish harvest. Estuaries are a type of coastal wetland. They slow down the effects of runoff by absorbing and partially filtering it before it reaches the ocean. They can absorb the force of ocean storms and prevent coastal erosion. They provide safe havens for migratory birds.

In 1987, the U.S. Clean Water Act was amended to form the National Estuary Program. The purpose of this program was to allow states to nominate estuaries of national significance and to request a management conference to develop a comprehensive estuary management plan. Since then, states and federal environmental agencies have generally given estuaries the same protection as wetlands.

From a sustainability point of view, estuary protection is very important. Species and genetic biodiversity flourish in these areas, and they help mitigate the spillover of pollutants from one ecosystem to another.

Groundwater

Lakes, rivers, and streams may exist wholly or partly underground. These water sources are called groundwater. Water percolates or seeps down through the soil into underground aquifers and streams. This recharges the aquifer with fresh water.

The rate of groundwater recharge is of keen importance to sustainability planners. Water is a basic system on which life depends, and it needs to be sustainable. Parts of much new ecosystem analysis include the recharge rate of the groundwater. If the recharge rate is too low, the water can become tainted with toxic pollutants. A concern of many environmentalists is that the recharge rate has been too low for too many years and that the environmentally degrading land use and industrial process of the past are seeping into the groundwater.

Hydrogeologists are currently mapping the course of groundwater. Many of the largest aquifers are known. The next step is to determine the present and future water quality of the ground water.

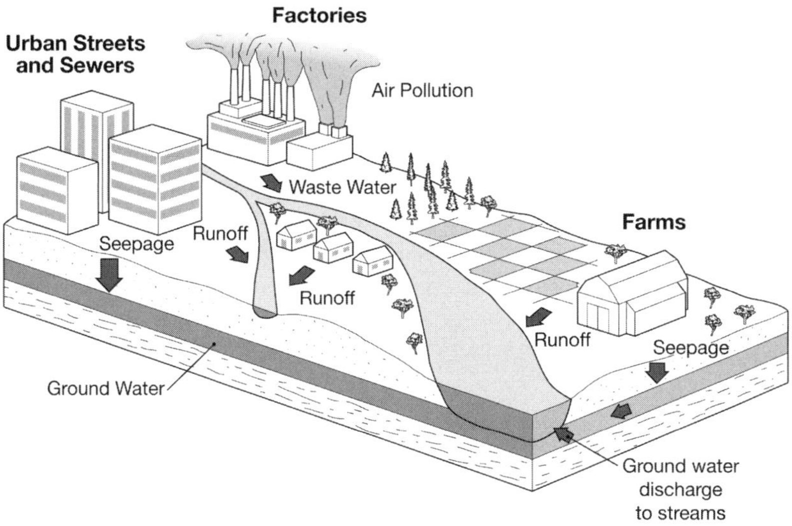

Figure 1.11 • Groundwater contamination can come from sources located on land such as factories, farms, and urban streets and sewers. Air pollution can also contribute to groundwater contamination as it falls to earth dissolved in precipitation. Illustrator: Jeff Dixon.

Runoff

Water flowing from land toward water is runoff water. This water carries with it many kinds of material. Dissolved material within the flow of this water is also called runoff. One of the contexts for runoff in sustainability relates to the intensification of agricultural usage of fertilizers and pesticides. These run off the land into rivers and eventually into the sea. For example, the death of the coral in the Great Barrier Reef of the east coast of Australia is due in part to excessive nitrogen and phosphates that is found in currently used agricultural products and processes. This type of runoff can cause "dead zones," or areas around the deltas of polluted rivers where no life exists. The number and size of dead zones have been increasing, and they are an area of study by environmentally based sustainability advocates.

Runoff also results from paving over land. The human landscapes of roads, streets, parking lots, airports, and malls are usually paved with hard surfaces. The water that would normally percolate into the ground and be slowly released into the ecosystem instead runs off the hard surface to wherever it is channeled. It is often channeled into public formal or informal sewer systems. Some sustainable communities are exploring alternative to nonporous surfaces.

Saving Water on the Roof

Other communities are exploring the idea of green roofs in the context of water conservation. Rainwater conservation is not a new idea. Many arid locations try to capture rain in various containment systems, both above and below ground. A green roof uses plants that grow on the roof to contain water. The roof must be waterproofed and able to repel roots. The green roof also needs to have a good drainage system. They are popular in Europe but just starting in some places in the United States. They can be modular or interlocking. The building structure of the roof itself must be strong enough to support the additional weight. In terms of water retention, green roofs can hold storm water and drain it more slowly. This helps prevent storm water sewer overflows. Many U.S. cities have consolidated sewer systems that overflow the waste treatment plants when storm water flows overtake their capacity. A green roof moderates the flow of water from rainstorms so that this does not occur and thus preserves fresh water. Storm water overflows from older urban consolidated sewer systems are a source of groundwater contamination. A green roof can also transfer water to the immediate environment through processes of evaporation and transpiration. This can have a cooling and cleaning effect on the air around the building. In most summer locations in the United States, green roofs can retain 70 to 90 percent of rain and between 25 and −40 percent in the winter, depending on the types of plants that are grown.

The city of Portland, Oregon, installed its first green roof in 1998. In 2008, about 8 acres of roofs in Portland were green. The city's goal is to increase this amount to 51 acres out of the 12, 400 acres of rooftops in Portland. The city calculates that this prevents about 50 percent of storm water runoff. The city offers an incentive program called Grey to Green Grants to help offset the higher cost of installation of green roofs. It is estimated that green roofs last twice as long as shingle roofs in Portland's rainy climate.

References

Hlavinek, Petr et al. 2009. *Risk Management of Water Supply and Sanitation Systems.* New York: Springer.

Kassim, Tarek A., and Kenneth J. Williamson. 2005. *Environmental Impact Assessment of Recycled Wastes on Surface and Ground Waters.* New York: Springer.

Mulamoottil, George, Edward A. McBean, and Frank Rovers. 1998. *Constructed Wetlands for the Treatment of Landfill Leachates.* Boca Raton, FL: CRC Press.

Rosemarin, Arno et al. 2008. *Pathways for Sustainable Sanitation: Achieving Millennium Development Goals.* London: IWA.

Tow, Philip, Ian Cooper, and Ian Partridge. 2009. *Rainfed Farming Systems.* New York: Springer.

Water Cycles

Water moves through the air and land in cycles depending on the climate and the landscape. Water interacts with the land and the air driven by thermal conditions. The pattern of these interactions can be observed as a repetitive cycle. Global, regional, and local water cycles occur continuously. Processes of precipitation, infiltration, runoff from impermeable surfaces, runoff generally, evaporation, and transpiration all contribute to the water cycle. Fresh water is reliant on ecosystems to filter it through physical, chemical, and biological filtration processes.

Watershed

Water flows from its sources to its endpoints from a variety of tributaries. Along such flows, water interacts with land and air in cycles. The distinct geographic area of such interactions from source to end is a single watershed. They are part of emerging measures or indicators of sustainability. The Earth Summit Agenda 21 puts adoption of an integrated, watershed-focused approach to water quality as a goal.

Some environmental groups balk at watershed programs because they see problems of effective environmental enforcement. In many nations, including the United States, the main focus of environmental regulation for water quality was on point sources of chemical discharges. However, nonpoint sources, unregulated industries and cities, and development in general contribute large amounts of pollution to the water, causing environmental degradation of natural systems.

Watersheds are part of many controversial issues of water use. As water quality and quantity become one of the first natural systems to erode, they become an environmental indicator. With climate change and the potential desertification of the equatorial tropics, watersheds will receive renewed attention. The use of watersheds is thought to increase accountability for actual water usage and water practices. The U.S. EPA allows people to surf their watershed.

In many places, agricultural industries consume the bulk of available water. In many urban areas in developing nations, water conservation is seldom practiced despite decreases in quality and quantity.

In 1996, the U.S. EPA categorized threats to aquatic species by watershed. They found at least one aquatic species to be at risk in 403 watersheds, between two and five species at risk in 745 watersheds, and more than five species at risk in 422 U.S. watersheds.

Wetlands

Land that is usually under water or soaked in water is called a wetland. Wetlands provide many ecosystem services that cannot be duplicated or replaced by any human-made system including cleaning of water polluted with nitrates and other runoff from urban and agricultural sources; protection of land from the destructive forces of floods, tides, and winds; and the protection of a variety of plant and animal species.

Wetlands are a flashpoint in environmental policy. Before the value of wetlands to ecosystems was known, they were considered swamps. To make these wetlands more habitable, many "swamps" were filled with dirt, rubble, garbage, and concrete. It is estimated that the state of Louisiana was 25 percent "swamps." Private and public policies of land development filled in about one-eighth of the state's "swamps." Swamps and other wetlands such as estuaries (see previously) have historically been the dumping ground for all types of residential, agricultural, shipping, and industrial wastes. Even if the wetlands are not the initial dumping ground of waste, they can become the site where waste becomes ecologically lodged, and sometimes toxic. Their value to the ecosystem and the legacy of environmental degradation highlights the preservation and expansion of wetlands in any environmentally based sustainability policy.

Wetlands are also considered part of the mitigation package for some of the possible changes in climate resulting from global warming. Wetlands can absorb enormous amounts of water, which helps to mitigate flooding. Flooding can come from many sources such as hurricanes, heavy rains, high ocean tides, dam breaks, landslides, and avalanches.

Overpaving natural wetlands creates runoff, which can contribute to the potential for flooding. Road and land development are the main sources of overpaving.

Reference

Mulamoottil, George, Edward A. McBean, and Frank Rovers. 1998. *Constructed Wetlands for the Treatment of Landfill Leachates.* Boca Raton, FL: CRC Press.

Ecosystem Values Become Apparent after the Ecosystem is Destroyed

The use and abuse of natural resources increase as population increases and as technological development increases. In the Aral Sea Basin of Central Asia, one of the largest inland lakes in the world became an abused natural resource. Over two-thirds of its entire volume was taken for irrigation of agriculture. When water is taken in this manner, it is called a "river diversion."

In all, 20 of the 24 fish indigenous to the lake are now extinct because of habitat reduction. In the 1950s, the commercial marine harvest was about 40,000 tons of fish per year. Many people also fished the river for food to live on, known as subsistence fishing. The marine harvest is now virtually nonexistent, because there are too few fish as a result of the river diversions of the lake. The wetlands surrounding the lake have decreased substantially and have caused a decrease in migratory waterfowl.

CHAPTER 3

Government and United Nations Involvement

ROLE OF GOVERNMENT IN SUSTAINABILITY

Governments have a role to play in organizing human activities. They control how and when business is conducted, what environmental resources to protect, and whether people are treated fairly. The challenge of sustainability to governments is to define its role in achieving environmental and ecological sustainability. What actions and values can a government foster to make sure that human economic activities and human communities live respectfully within the limits of the webs of life supporting all life on this planet?

Public Policy and Sustainable Development

All kinds of government, including monarchies, democracies, republics, and colonies, have public policies. The main intent of all these policies is generally to protect the health, safety, order, and welfare of society. Many policies are also designed to facilitate business and trade. Sometimes governments are not representative of all the people and develop policies to protect select constituencies. If these constituencies create degrading environmental impacts, then confrontation and change between these groups and sustainable development advocates develop. At other times, the state or nation and its government do not reach all the people or ecosystems in a given geographic region. For example, in many parts of the equatorial rainforests, there are regions virtually ungoverned by any nation, but all geographic regions are claimed by nations. Public policies about sustainable development in these regions of known biodiversity and rapid and controversial deforestation face challenges when there is no governmental public policy that protects natural systems on which all life depends.

Public policies are those policies and practices developed by the chief executive or executive group of government, legislation, law, court decisions, and administrative agencies. The government pursues different policies through rules, regulations, taxes, and subsidies. These polices have the force of law and are the way in which many sustainability advocates pursue social change. Public policies and personal practices need to change in many places to begin the process of sustainable development. To the extent public policies around sustainable development affect personal patterns of consumption, they can be effective in reducing environmental impacts.

Public policy is by its nature political. Environmental values, power struggles, and controversy mark early sustainable development policy. Public policies determine the distribution of benefits and burdens that affect human behavior. Policy issues of regulatory efficiency, how costs and benefits are measured, the participation and access of stakeholders in the process, and the distributional impacts of the policy all become necessary for policies of sustainable development. All these policy issues are also dependent on values, and sustainable development is dependent on meaningfully incorporating environmental values. Most policies are evaluated by the indicators of value in that society. Environmental standards are new to public policy to the extent needed for sustainable development. In the United States and most other industrialized nations, economic values are indicators of value that occur in the absence of developed or enforced environmental standards. Only those costs that are measurable are emphasized when developing public policy in the United States.

Intangible costs of cumulative impacts on public health and on ecological integrity are generally not counted because they are too intangible. This is problematical under sustainable development because not only do the effects on the environment accumulate, they can do so without predictable impacts on the environment. As one scholar notes:

> Whereas some impacts pass away soon after their termination, leaving few if any marks of their occurrence, the effects of other activities may be cumulative. This is obviously so in the case of certain physical events but may also be characteristic of more purely sociocultural impacts. . . .

The effect of accumulation may not always conform to the expectations derived from simple arithmetic. For example, the impact of one offshore support vessel working out of a small harbor might be slight; the impact of a second might double the impact but still be slight. After some increase in the number of vessels, however, qualitative changes may commence. New docks, new fuel delivery routes, changes in the proportions of persons employed in the hydrocarbon versus fishing industries and their consequent places in the local economies political arena could result.

Roy A. Rappaport, 1994

Pain, suffering, the beauty of nature, the costs to future generations, and the value of irreplaceable natural systems are also intangible and generally not captured by traditional public policy analysis. Freedom is also highly valued, not captured by traditional cost benefit analyses, and transcends governments. As noted by Alexis de Tocqueville in *Democracy in America:*

> Freedom has appeared in the world at different times and under various forms; it has not been exclusively bound to any social condition, and it is not confined to democracies. Freedom cannot, therefore, form the distinguishing characteristic of the democratizing ages. The peculiar and preponderating fact which marks those ages as its own is the equality of conditions; the ruling passion of men in those periods is the love of this equality.

Alexis de Tocqueville, 1863

The treatment of all people equally resonates with equity aspects of sustainability and with sustainability generally. Unequal treatment of people in the implementation of a young environmental regulatory protocol results in pockets of waste build up. The unequal distribution of environmental benefits and burdens sparked the U.S. environmental justice movement. In the United States "freedom" is a strong value often associated with ownership of private property. Individual property owners, including large industrial land users, are theoretically "free" to use their land, water, and air any way they want as long as there is no direct danger to the owner or to others and their private property. The problem develops when the land uses result in environmentally degrading impacts that threaten public health or ecological well-being. As the preceding quotation observes, however, it is the love of equality of all people that is most characteristic of freedom, not the ownership of land. Public policies at local, state, federal, ecoregionally, and globally of sustainable development and private property face strong stresses as the evidence of our impact on global natural systems forces a reexamination and reprioritization of values. An area that sees many of these stresses first is the area of public policy analyses.

References

De Tocqueville, Alexis. 1904. *Democracy in America.* Translated by Henry Reeve. New York: D. Appleton and Co.

Houck, Oliver. 2009. *Taking Back Eden: Eight Cases That Changed the World.* Washington, DC: Island Press.

Rappaport, Roy A. 1994. "Human Environment and the Notion of Impact." In *Who Pays the Price? The Sociocultural Context of Environmental Crisis,* ed. Barbara Rose Johnston, 167. Washington, DC: Island Press.

PUBLIC POLICY ANALYSIS AND SUSTAINABLE DEVELOPMENT

Public policy analyses in Western industrialized countries like the United States tend to be based in economics only. They measure economic impacts and related behavior changes of some public policies. Many public policies are political, serving to meet the needs or acquiesce to the demands of large constituencies or stakeholders. For example, many natural resources owned by the U.S. federal government are leased to private corporations for a variety of environmentally degrading uses such as grazing, logging, and mining. In what many sustainable development advocates describe as perverse, some very damaging environmental activities are subsidized by the government. In terms of environmental policy development, the foundational economic value often misses or undervalues ecosystem services or environmental impacts. Public policy analysis that relies on economic growth as a foundational value also often misses or undervalues ecosystem services or environmental impacts.

Nonetheless, sustainable development will be subject to traditional public policy analyses. Whether these models can be adapted to longer term and more ecologically based sustainable development is unknown. From an ecologically based sustainable development perspective, the answer often depends on the value of the environment, the value of getting information about ecological impacts, and the scale of time. If an action or project threatens to destroy systems on which all life depends, then sustainable development would require a precautionary investigation into ecological impacts. Can traditional public policy analyses capture these aspects of sustainable development?

Cost-Benefit Analysis

Cost-benefit analyses are often used to set priorities within and between public policies. The main goal of this analyses is to develop a public policy that most efficiently uses resources to solve the given public policy problem. Cost-benefit analysis places a monetary value on the consequences of a given public policy. It can be applied to projects and policies. Both costs and benefits must be expressed, or valued, in monetary terms. Many economists call this "getting the prices right" for environmental or ecosystem services. It is a very controversial analysis in environmental circles.

Under sustainable development, many aspects of nature are priceless, or at least outside the realm of any one generation paying for it. As noted by scholars:

> The basic problem with narrow economic analysis of health and environmental protection is that human life, health, and nature cannot be described meaningfully in monetary terms: they are priceless . . .
>
> There is no reason to think that the right answers will emerge from the strange process of assigning dollar values to human life, human health, and nature itself, and then crunching the numbers . . . formal cost benefit analysis often hurts more than it helps.

But for most people, there are matters of rights and principles that are beyond economic calculation. Setting boundaries of the market helps to define who we are, how we want to live, and what we believe in. There are many activities that are not allowed at any price.

Frank Ackerman, 2004 Another area of cost-benefit analysis is human health risk assessment in environmental policy making. Cost-benefit analyses are seldom applied to ecological risk assessments, which are needed for sustainable development. In terms of human health risk assessment, the application of cost-benefit analyses results in the application of the concept of marginal utility to human life and environmental protection. Marginal utility means that the costs have outweighed the benefits of a given activity. When applied to human risk assessment, it means that the cost of removing the last 0 to 50 percent of a given pollutant may be more than the value assigned to the human lives affected by the chemical release.

A market–based public policy tends to accept the marginal utility of human life. In a private sector economy, it is difficult to "force" a private business to take a loss for the advancement of social policy. Many communities, environmentalists, and others find allowing any marginal utility an affront to the value of human life. Regulatory policies of sustainable development need to consider this belief both domestically and globally. Some economists, such as Lawrence Summers, director of the White House's National Economic Council, have advocated for moving heavy polluting industries into the poorest, least regulated nations under marginal utility concepts. *See also* **Volume 1, Chapter 4: Risk Assessment.**

References

Ackerman, Frank, and Lisa Heinzerling. 2004. *Priceless: On Knowing the Price of Everything and the Value of Nothing.* New York: New Press.

Sachs, Jeffery. 2006. *The End of Poverty: Economic Possibilities for Our Time.* New York: Penguin.

Stiglitz, Joseph E. 2003. *Globalization and Its Discontents.* New York: W. W. Norton.

Wolf, Martin. 2004. *Why Globalization Works.* New Haven, CT: Yale University Press.

Cost Utility Analysis

One method used in U.S. health care policy is cost utility analysis. This is effectively a modification of the cost effectiveness analysis. The modification is designed to focus on the quality of life issues around health care. The main measure of outcomes is a "quality adjusted life year" or QALY. This measures the number of years of life gained by a health policy adjusted for the quality of the extra years of life. Cost utility analyses are used when the quality of life is the main outcome of the policy, when comparing morbidity and mortality, and when comparing cost utility analyses with other similar analyses. Quality of life measures are important for sustainable development. Humans are parts of ecosystems and biomes, and degradation of ecosystem health usually results in degradation of human health.

Cost-Effectiveness Analysis

A more modern tool of public policy analyses is the cost-effectiveness analysis (CEA). CEA is a technique for comparing the relative value of various policies. In its most common form, a new policy is compared with current policy, which is sometimes called the "low-cost alternative." Although there are variations depending on agencies and their mission, most follow the following process. Each one of these stages in the process can become quite involved. First the public policy problem is framed. At this stage a number of policy options are considered and brainstormed. Next, appropriate outcome measures for the success of the policy are developed. The next stage in the process identifies the best public policy choice and the costs of that outcome. Sometimes the

Sustainability Impact Assessment: Tools for Environmental, Social, and Economic Effects of Multifunctional Land Use in European Regions

The SENSOR project is an ambitious initiative bringing together researchers from more than 30 institutes in 15 European countries. SENSOR has described itself in the following terms:

> Sustainability of land use in European regions is a central point of policy and management decisions at different levels of governance. Implementation of European policies designed to promote and protect multifunctional land use requires the urgent development of robust tools for the assessment of different scenarios' impacts on the environmental and socio-economic sustainability. The technical objective of SENSOR is to build, validate and implement sustainability impact assessment tools (SIAT), including databases and spatial reference frameworks for the analysis of land and human resources in the context of agricultural, regional and environmental policies. The scientific challenge is to establish relationships between different environmental and socio-economic processes as characterized by indicators considered to be quantitative measures of sustainability. Scenario techniques will be used within an integrated modeling framework, reflecting various aspects of multifunctionality and their interactions. The focus will be on European sensitive regions, particularly those in accession countries, since accession poses significant questions for policy makers regarding the socio-economic and environmental effect of existing and proposed land use policies. SIAT will utilize the statistical and spatial data continuously collected by European and regional agencies. SENSOR will deliver novel solutions for integrated modeling, spatial and temporal scaling and aggregation of data, selection of indicators, database management, analysis and prediction of trends, education and implementation. SIAT will be made available to decision makers at the European and regional level, providing user-friendly interfaces and scientifically sound procedures for the assessment of environmental and monetary responses of policy options.

next stages of these processes develop decision trees to analyze public policy alternatives.

If a policy is "cost-effective," it means that the new policy is a good economic value. Cost-effective does not mean that the public policy promotes environmental protection or that environmental standards are developed. Cost-effective public policy analyses require a value judgment. CEAs used in U.S. public policy is a specific type of economic analyses in which all costs are related to a single, common effect, or public policy goal. It allows decision makers to compare different policies in similar terms and scales. It evaluates public policy options quantitatively within the confines of a defined model. It requires measurable costs and outcomes, and tends to ignore intangible measures. Some early sustainable development land use policies are applying CEA to early results.

Reference

Levin, Henry M., and Patrick J. McEwan. 2000. *Cost Effectiveness Analysis: Methods and Applications.* Thousand Oaks, CA: Sage Publications.

Changing Goals and Rules

Governments shape cultures by articulating certain values in the form of goals, and setting the rules by which such goals are to be accomplished. Legislation and regulation shape and define such rules. Governmental leaders often define and distinguish themselves from others by articulating such values and rules.

Standard Setting

Legislation and regulation often set standards for economic and social activity. In terms of environmental protection, current law has mostly set standards in terms of allowing some forms of pollution and waste as a cost of economic and social activity. Setting the best amount of waste/pollution to economic benefit (sometimes called an optimal amount) has fostered a sense of the right to pollute in the form of permit or permission to pollute instead of setting standards (and goals) aimed at elimination of waste and pollution. Government has a powerful role to play in setting standards that challenge human activities to prevent and eliminate waste and pollution.

ARE THE STANDARDS FOR ENVIRONMENTAL POLICY USEFUL IN THE DEVELOPMENT OF STANDARDS FOR SUSTAINABILITY?

The standards that are set for the regulation of the environment are generally designed to protect the public health, safety, and welfare, and to protect the environment. Many federal and state agencies set standards to help develop rules and regulations, and to determine how these rules and regulations are enforced. The U.S. Environmental Protection Agency (EPA) first began developing standards for clean air, clean water, and treatment of solid wastes in the early 1970s. The process for developing standards for environmental protection is new, dynamic, and controversial. In most nations only those industries with specific emissions and pollutants that rise above a given threshold are regulated. Whether an industry exceeds a given threshold is often determined by the industry itself. Many environmentalists and communities complain that industry is not honestly self-regulating, that the thresholds are set too high for true environmental protection, and that the list of chemicals that are regulated is not sufficient to protect the environment. If these complaints are true, then the current regulatory framework for environmental protection may not be a good foundation for a policy of sustainability that would account for all environmental impacts.

There are many groups in society other than small or select industries that are not environmentally regulated. These can be towns, nonprofit organizations, schools, farms, and individual households. These groups may engage in activities that have environmental impacts that could affect sustainability.

WHAT ARE THE STANDARDS?

One group of standards for environmental protection is called a "specification" standard. With these types of standards the federal agency dictates exactly how the regulated group will achieve a specific environmental result. This is opposed to "performance" standards. Performance standards are set by the regulatory agency, and the regulated group chooses the performance to meet that standard. The difference between the two approaches to standard setting is the degree of independent decision-making power that the regulated group has to meet a given environmental goal. Industries prefer to have performance standards so that they can control costs.

Another group of standards are directly related to a reference point of human health or environmental protection. These are called "health-based" or "risk-based" standards. In these types of standards, the agency determines how much the public or the environment can be exposed to before it is unsafe. In this determination, calculations are made about the carrying capacity of a part of the environment. This aspect of current environmental standard setting could prove useful for future sustainability regulations. Also in these calculations is a determination of how much risk there is to the public. Currently, this information examines the death rate per number of people exposed to the regulated chemical. This process of risk assessment is controversial. Another set of standards used to develop rules and regulations to protect the environment is called "technology-based" standards. These standards are based on a determination of what current technology in pollution control and abatement will allow the industry to achieve. Environmentalists challenge this standard-setting mechanism because it is independent of whether the environment is protected. Related to this standard-setting mechanism is the "technology forcing" type of regulation. Here the agency develops a standard that is achievable only by technology not currently available in the regulated group.

The current set of environmental rules and regulations uses a combination of all these standard-setting techniques. Each technique has serious flaws in its current form in terms of environmental and human health protection. They do not include all environmental impacts, ignore the accumulation of impacts, fail to account for global conditions of the environment, and exclude community concerns.

Reference
Collin, Robert W. 2006. *The Environmental Protection Agency: Cleaning Up America's Act.* Westport, CT: Greenwood Press.

Risk Assessment and Management

Chapter 2 defines risk assessment and management. Government often requires a risk assessment in many areas of environmental and developmental activity. These studies are used to set priorities and determine funding priorities. Often, risk assessment is done as a matter of expert assessment. Risk perception, however, is a developing aspect of this science, and a commonly held perception of environmental and ecological risk can provide an important common platform for environmental action that is missing in current controversies. Government has an important stake in developing such common ground as a basis for legislation and regulation for sustainability, especially in changing assumptions of permissions to pollute toward assumptions of eliminating all forms of waste and pollution.

RISK ASSESSMENT APPLIED TO SUSTAINABILITY

The type of risk assessment most likely to be applied comes from environmental and ecological risk assessments. Environmental risk assessments tend to look at the risk to one species or one place. They have tended to focus on human health. Ecological risk assessments were begun in 1990 at the U.S. EPA and are primarily used in the cleanup of Superfund waste sites. They are designed to examine adverse effects on life forms within ecosystems. They are complex and controversial. Issues of cause and effect within an ecosystem can depend on the length of time in which the effects are measured. A short time span in an ecological risk assessment can effectively negate any scientific finding of cause and effect between an environmental stressor and an adverse ecological impact. Controversies also swirl around which effects are "adverse."

One risk assessment policy mandated by the U.S. EPA and based on sustainability principles is the Endangered Species Act, which began as the Endangered Species Preservation Act, passed in 1966. This early act was motivated by the concern that rapid growth in technology and development could render certain species extinct. Programmatically, its impact was small because it was limited to acquisition of habitat for threatened species. In 1969, the act was changed to the Endangered Species Conservation Act. This expanded the programmatic coverage of the law to include species threatened with worldwide extinction. Because the risk of loss of species and extinction is itself an irreparable risk for future generations, however, this act was changed to the Endangered Species Act of 1973. The agencies charged with implementing this act are the Fish and Wildlife Service in the Department of Interior and the National Marine Fisheries Service of the National Oceanic and Atmospheric Administration in the Department of Commerce. This act mandates two types of risk control: risk assessment and risk management. The agencies are required to evaluate the risk of species extinction when the species becomes endangered. Endangered means that it is in danger of extinction throughout all or a significant portion of its habitat. Endangered also means that it is threatened with extinction, which means that it is likely to become an endangered species within the foreseeable future throughout all or a significant portion of its range. If a species is listed as endangered, then the federal government must begin a risk-management process to prevent extinction. The Endangered Species Act has been amended several times since 1973, and has faced substantial controversies. It is considered one of the strongest environmental laws in the United States and directly confronts pivotal human values of private property, planned natural resource use, and capitalism. Nonetheless, it still uses risk assessment to try to preserve species from extinction so that future generations may appreciate them.

Risk assessments concerned with sustainability face a serious problem of uncertainty. Scientific models of causality, lack of knowledge, and the need for private and public action all push this uncertainty to the forefront. This is particularly acute at the international level in risk assessments of global climate change. In 1988, the United Nations formed the Intergovernmental Panel on Climate Change to provide the scientific basis of risk from human-caused climate change. The panel has published four reports since its inception and each report discusses the uncertainty around risk formulation in great detail. The first 1990 report concluded that global climate changes were observable, but there was too much uncertainty to conclude whether it was human caused. The 1995 report concluded that there was a discernible amount of human influence on global climate change, but that there was still great uncertainty about it. In its 2007 report, the panel decreased the uncertainty by concluding that most of the increase in global warming was "very likely" caused by increases in human-produced greenhouse gases. There is still great uncertainty over how climate change will affect a particular location, but less uncertainty over whether the ice caps are melting and can cause a rise in ocean levels. Risk assessments used for sustainability may decrease some uncertainty and retain other areas of uncertainty. Financial institutions have developed their own principles for evaluating environmental risk in terms of financing large projects such as the

Equator Principle. (See further discussion throughout volume 2 of this encyclopedia, which focuses on business and economics.)

Risk assessments are the primary vehicle scientists use to evaluate threats to the environment. They are used by governments and international financial institutions to develop policies and projects. They will be used much more frequently to determine if there are threats to natural systems on which future life depends. These risk assessments may be used to determine whether to apply the precautionary principle, which has a lower standard of proof than current scientific cause-and-effect models. *See also* **Volume 1, Chapter 2: Waste, Pollution, and Toxic Substances; Volume 1, Chapter 5: Risk Analysis.**

References

Calow, Peter. 1998. *Handbook of Environmental Risk Assessment and Management.* Lanham, MD: Government Institutes.

Robson, Mark G., and William E. Toscano. 2007. *Risk Assessment for Environmental Health.* Hoboken, NJ: Jossey-Bass.

Government as Consumer: Contracting

Governments exercise unique power to influence voluntary private marketplace transactions without the need for legislation or regulation. This power is often exercised through government acting in its role as a purchaser of products. The specification of sustainable products and materials in government procurement can create a market for technology and materials that stimulate investment and innovation. Without such a stimulus, the costs of innovation might deter changes in technology and materials in many sectors of the economy. Many levels of government, including local, state, and federal, are focusing on procurement of sustainable goods and services. Some federal agencies are going beyond procurement approaches to sustainability to multistakeholder cross-media, problem-solving approaches.

The U.S. Department of Defense and the Environmental Council of the States have formed a sustainability work group. This forum provides a venue for discussion by U.S. EPA officials assigned to defend facilities and the Environmental Council of States. Federal facilities management deals with complex and sometimes controversial environmental issues such as the cleanup of contaminated sites. Many defense facilities are sites for hazardous and toxic wastes that can migrate from the site. The charter of the Environmental Council of States states:

Environmental Council of the States [ECOS] hereby establishes The DOD [Department of Defense] Sustainability Work Group Which will, under the Waste Committee:

- Serve as a focal point within ECOS for dialogue on sustainability issues with regards to DOD installations;

- Serve as a forum in which ECOS and DOD can discuss and address issues and share examples of success related to sustainability at DOD facilities.

- Conduct such research on these issues as the Work Group deems necessary;

- Share information with ECOS members regarding programs and activities related to sustainability at DOD installations;

- Develop proposed ECOS policy positions on DOD sustainability issues in cooperation with the Waste Committee.

The sustainability work group has two task groups. The first task group is the emerging contaminants task group. This group is charged with studying and developing approaches to emerging contaminants. These are materials that are released into the environment that may potentially affect human health, natural resources, or ecological systems and for which no scientific risks are known. This is an important step in sustainable development because it allows the study of environmental impacts that could threaten natural systems on which future life depends without waiting for scientific pronouncements of risk, which can be too late to inexpensively mitigate once they are proven. The other task group examines compatible land uses and sustainable development in and around defense sites. These can be extremely complex issues. Both these groups have generated a number of research papers that are publicly available.

Federal facilities management of high environmental impact facilities represents the leading edge of public policy around sustainable development. The federal government is free to require stricter standards in the goods and services it procures, and in the plans it makes for the future. Because of these factors it can lead the way for state and local governments, and for broader public policies that reach the public and the global community.

Reference

Fort Carson Sustainability and Environmental Management System Plan, www.fedcenter. gov/_kd/Items/actions.cfm?action=Show&item_id=10217&destination=ShowItem.

Taxes and Subsidies

Taxes and subsidies are additional ways in which government significantly influences private, voluntary choices. Subsidies are payments that reward conduct deemed desirable. Taxes require payments from individuals or businesses. One use of taxes is as disincentives for conduct that government wants to slow or eliminate. Taxes can attach negative consequences to a particular choice, making other options more attractive. Subsidies can attach positive consequences to a particular choice, making that option more likely to be chosen. These tools of government can rationally be applied in a broad variety of ways to deter unsustainable behaviors and to encourage sustainable behaviors. Our history, and expectations derived from policies of the past, however, have resulted in taxes and subsidies that work in perverse ways. Expectations that are now entrenched may tax desirable things, and subsidize undesirable things and behaviors.

It is estimated that government subsidies worldwide provide about $500 billion a year for environmentally destructive activities such as logging, fishing, and mining. Still, if these subsidies were eliminated, some of these activities would continue. Subsidies are ultimately paid for by taxpayers who include private citizens and business entities. If these subsidies were eliminated, some estimate that taxes would decline by about 7 percent. The costs of environmental and health damage caused by subsidizing environmentally destructive activities would be saved; however, most of these estimations do not include the value of ecosystem services.

The extent of subsidies is large and often political. Economic development through natural resource use often occurs when nations

Figure 1.12 • Incumbent energy resources are subsidized by government. Oil rigs sit in Lake Maracaibo in western Venezuela, South America's richest oil-producing area, on November 17, 1970. While oil revenues are the backbone of Venezuela's economy, supplying 70 percent of government income, the black gold has created a host of problems. Thousands of acres of the nation's land lie fallow while Venezuela has to import foodstuffs it could produce. AP Photo.

subsidize business organizations. Petrochemical industries are heavily subsidized, especially for research and site development. The British Conservative Party estimated that for every pound sterling spent subsidizing renewable fuels, four pounds were spent on subsidizing oil and petrochemicals.

Developing nations subsidize their natural resources to develop capital for economic growth. Research shows that the more a developing nation relied on natural resource exports, the slower the economic growth and the lower the average income. Many subsidies were intended for smaller scale operations of logging or mining, but have been exploited by large multinational corporations.

Subsidies can occur based on the way the government doles out land. The United States has created mineral subsidies by giving away or selling very inexpensively federal land with about $242 billion of gold, silver, and other minerals since 1873. This is due to the Mining Law of 1872. Designed to encourage Western settlement, this law allows anyone, including foreign corporations, to find minerals and patent them at the 1872 price. Patenting means they get to use the land as long as they are mining it. There are no royalties paid on their patents, or minerals like gold, silver, oil, and gas. Currently there is a moratorium on the granting of patents. There have been moratoriums declared before and they have been defeated in the courts. There is also no provision for any cleanup and mining leaves behind many toxic and environmentally hazardous materials. On top of that, mining has numerous tax deductions for exploration and mine closing. The Mining Law of 1872 is an example of one of the most pernicious and nonsustainable subsidies in the United States. It is expensive, dangerous to the environment, and controversial. Some countries are starting to phase out subsidies of nonrenewable energy sources. For example, France, Belgium, and Japan are phasing out subsidies for domestic hard coal production.

Attacking subsidies is difficult because they are often politically entrenched, with some people relying on them for jobs. As an example, there is a congressional fight to eliminate subsidizing the building of a logging road in the Tongass area of Alaska. Tongass National Park is a large area with great biodiversity and several intact rainforest ecosystems. Subsidizing the logging road provides construction jobs as well as a road in a desolate place that can be used for all types of economic development other than logging. In June 2007, the U.S. House of Representatives passed the Chabot/Andrews amendment to eliminate taxpayer subsidies for logging road construction. It was passed in the House by a large bipartisan majority but dropped when the bill went into conference committee in the U.S. Senate. The bill is being reintroduced in 2009 as part of the annual Department of Interior Appropriations bill. The bill's sponsors charge that the U.S. Forest Service already wasted $48 million in subsidies to clear-cut a U.S. rainforest. They contend that future subsidies would increase to $1.2 billion, without considering environmental degradation, cleanup, or loss of ecosystem services.

The list of subsidies in the United States is long, but here are some examples and current cost estimates for another country, Australia. In 2007, Australia provided $1.1 million in subsidies for company cars in the business sector. Sustainable devilment advocates point out that governments can provide incentives to companies to have low carbon transportation options such as bicycles, alternative fueled cars, and public transit passes. Airplane gas taxes are much lower than for the rest of gas consumers, creating a subsidy of about $830 million. Airplanes create substantial amounts of air pollution via greenhouse gases and are a rapidly growing source and conduit for economic development. Corporate air travel is a large contributor to this economic growth. Australia had a hole in its ozone layer in the early 1990s that caused great concern. Sustainable development advocates think that emissions could be reduced by video conferencing, Internet use, and better trip planning.

Subsidies for environmentally degrading activities are a threat to sustainable development when they threaten systems of nature on which future life depends. Some subsidies began when the human population was smaller, the technology less powerful, and the methods of resource acquisition more gentle. Now that the human population has grown and is growing faster, and for the first time the industrialized methods of natural resource acquisition (mining, logging, ranching) threaten whole ecosystems, the threat to sustainable development is larger. Now that global markets pursue natural resources wherever they can find them with the latest technology and the least environmental enforcement, subsidies for older activities are a problem. ***See also* Volume 2, Chapter 4: Resource Subsidies.**

References

Freidman, Thomas. 2008. *Hot, Flat, and Crowded: Why We Need a Green Revolution and How It Can Renew America.* New York: Farrar, Straus and Giroux.

Meyers, Norman, and Jennifer Kent. 2001. *Perverse Subsidies: How Tax Dollars Harm the Environment and the Economy.* Washington, DC: Island Press.

Roodman, David Malin. 1996. *Paying the Piper: Subsidies, Politics, and the Environment.* Washington, DC: Worldwatch Institute.

Enforcement of Environmental Standards

One of the most obvious ways that government operates to achieve a policy of any sort is to set a standard of behavior. That standard may then become the basis of enforcement activities by others, government itself, or other private parties. Enforcement of a standard can be accomplished by coercive means such as punishments. Compliance with standards can also be accomplished by noncoercive measures such as incentives and compliance assistance.

Enforcement theories have different theoretical bases. Most environmental enforcement theories are to gently induce compliance with environmental standards. Other theories of enforcement include deterrence, retribution, and rehabilitation. Deterrence-based theories of enforcement include both general and specific deterrence. Specific deterrence theory would be a punishment for a violator of an environmental law that is specific to deter future violations. General deterrence theory is a punishment for a violator of an environmental law that is meant to serve as an example to others so that if they violate the law they will also be punished. General deterrence enforcement punishments must be known to the future potential violators in order to be effective. Retribution enforcement theories are meant to punish the wrongdoer for his crime. Rehabilitation-based theories of enforcement are meant to rehabilitate, or train, the wrongdoer about the reason why the act was wrong. If sustainability grows as a stronger social policy, these additional theories of environmental enforcement may be included.

Empowering Community

Government can also use its power to convene forums and discussions, and facilitate those meetings with technical assistance and other types of logistical support. These dialogues may result in the knowledge and ability of communities to become more active in their interactions with neighbors and with their local ecologies in ways that support the goals of sustainability.

References

Blewitt, John. 2008. *Community Development, Empowerment and Sustainable Development.* Devon, UK: Green Books.

Kasemir, Bernd, and Jill Jager. 2003. *Public Participation in Sustainability Science: A Handbook.* Cambridge, UK: Cambridge University Press.

Maida, Carl A. 2007. *Sustainability and Communities of Place.* New York: Berhahn Books.

WATERSHED VOLUNTARY LOCAL EFFORTS: WATERSHED COMMITTEES

Most states have developed watershed councils that cover the entire state. These councils are generally made up of volunteers and have little enforcement power. More and more citizens are concerned with the quality and quantity of their water resources. They are concerned that irreparable damage is being done and seek to become involved. They are creating their own organizations that monitor and clean up watersheds in their area. They work to restore rivers, streams, wetlands, lakes, beaches, and estuaries. More than 2,300 such organizations currently exist in the United States. They get training in water quality monitoring, develop baseline data, watch for polluting activities, and serve as stewards.

The U.S. EPA encourages these activities and will help organizations in these efforts. They provide a Web site that enable readers to surf their watershed: cfpub.epa.gov/surf/locate/index.cfm. ***See also* Volume 3, Chapter 2: Public Participation.**

References

Buckingham, Susan, and Kate Theobald. 2003. *Local Environmental Sustainability*. Boca Raton, FL: CRC Publishing.

Curwell, S. R., Mark Deakin, and Martin Symes. 2005. *Sustainable Urban Development*. London, UK: Taylor and Francis.

Ford, Andrew. 2009. *Modeling the Environment*, 2nd ed. Washington, DC: Island Press.

Monto, Mani, L. S. Ganesh, and Koshy Varghese. 2005. *Sustainability and Human Settlements: Fundamental Issues, Modeling and Simulations*. Thousand Oaks, CA: Sage Publications.

BARRIERS TO GOVERNMENTAL ROLES IN SUSTAINABILITY

Governments do not control all environmental actions. In many nations the role of the government is limited because of the lack of resources available. Many of these governments seek to increase the quality of life of their citizens no matter what the cost to the environment. In other nations the role of government is limited because forces outside the government are stronger. Some nations are in a state of civil unrest or war. They, too, are often without resources. Without adequate resources to study the environment and know ecological baselines, many governments cannot achieve sustainable development and could irreparably damage their environment.

Governments with resources may view the environment as a political issue and therefore subject to political negotiation and compromise. Some reject the notion of sustainability as a liberal issue that contradicts their understanding of the environment as a source of unlimited growth. Some governments are limited by their use and application of science. Governments that politically compromise the environment for their short-term economic gain also do not study the environment enough to know its ecological baselines and carrying capacities, and may risk irreparable damage to their environments for future generations.

Other governments have value structures that limit their approach to the environment. Private property values as a basis of strongly held liberty values may impinge on governments that seek to know the ecological carrying capacity in areas that transcend the boundaries of private property ownership. Boundaries of private and public property along with borders of nations blur the analysis of natural systems. Natural systems respond to other aspects of their ecosystem, not human-imposed boundaries.

Borders and Jurisdiction

The government of a country can exercise power over its territorial areas defined by its physical borders. These borders may not have a relationship to ecosystems or other environmental features. Instead, they may reflect other facts of history without regard for the geographical or cultural features of the place. Ecosystems and their features, like watersheds or mountain ranges, often cross borders between countries.

In some continents, such as Africa, North America, and South America, other nations came and colonized them. Many colonizing nations sought natural resources such as salt, gold, and oil and divided the land into parcels to efficiently retrieve these resources. In the process many indigenous people were killed and relocated, sometimes intentionally and sometimes unintentionally when diseases were introduced for which they had no resistance. The land was divided by borders that were defended and guarded. Some areas with known natural resources and viable transit routes were heavily exploited. Other areas with few natural resources and nonviable transit routes were often left alone. This global process of colonization has left ecosystem impacts that may affect modern efforts of sustainability.

In a modern political state, pollution can occur within one state or nation and be carried into another nation or state. Airborne pollution can travel long distances, such as in the case of acid rain. When pollution travels it can enter other parts of the ecosystem such as leaving the air to fall as rain onto the ground, and from there it can move into plants and animals. Environmental protection systems based on nations or states can facilitate simply moving regulated pollution into unregulated areas. Nations or states seeking to attract economic growth through industrialization may do so by environmental deregulation or lack of real enforcement of environmental rules. This is a process known as "race to the bottom."

Border and property boundaries that do not reflect ecological systems in nations that have little or incomplete environmental regulation pose a significant challenge to sustainable development. A foundational component of the emphasis on land division and human ownership of land is private property.

Reference

Just, Richard E., and Sinaia Netanyahu. 1998. *Conflict and Cooperation on Trans-Boundary Water Resources.* New York: Springer.

Ownership of the Environment: Private Property

Ownership of property gives individual owners the power to determine what choices to make regarding that property in some cultures. This right may exist without knowledge of, or regard for, the consequences of that individual choice for the ecology of the place. To the extent that government ensures such absolute expectations of rights to property, sustainable choices for environment and for community may be frustrated.

In the United States, 3 percent of the land is owned by industries and 2 percent is owned by private residences. Overall, in the United

States there are about 2.1 billion acres, which is about 7.7 acres per person, 14 times higher than the rest of the world and 3 times higher than any other industrialized nation. About 16 percent of that land is occupied by 341 metropolitan statistical areas. The federal government is a large landowner, owning 96 percent of Alaska, 86 percent of Nevada, 66 percent of Utah, 63 percent of Idaho, and 52 percent of Oregon, as well as substantial acreage in other states. This represents about 730 million acres, or about one-third of the total land and water area of the United States. The ability of the state to control the use of land, whether public or private, leased, owned, or held in trust, is important for any policy of sustainability.

WHERE DID THE VALUE OF PRIVATE PROPERTY ORIGINATE?

Where did the concept of "private property" originate? Understanding the roots of this value is important for sustainability because the settled expectation of private property owners may be forced to change. Monarchs (kings, queens) often claim lineage with God as the basis for their legitimacy of their power over their people and the land they rule. For many of the first U.S. colonialists, the concept of private property came from John Locke. Locke made "natural law" arguments for the creation of private property. He wrote that God, by making humans in his image, gave value to nature. Nature, the environment, had value to the extent humans labor gave it value. In early U.S. history, only white men over the age of 21 could own land and vote, and sometimes they had to own land just to register to vote. As industrialization and capitalism grew with the fledgling U.S. democracy, it became clear that the laborer did not own the land she labored on.

Land has many other values that are not easily valued in terms of the price of a piece of real estate. Some deep ecologists feel that to even put a price on land devalues it, and that humans are caretakers of the land for the environment and for future generations. Other values tied to land can include spiritual, cultural, historical, environmental, and sacred. Land may be valued as a place, for its human and community habitability and as an integral part of a system, for the strength of the ecosystem, its healthy diversity, and its resilience. These values are not necessarily reflected in market value.

Private property rights in land quickly reduced land values to the economic value of it or its products. The danger of this in terms of sustainability can be seen in many examples. It may be that a community dies while a market thrives, especially in a modern era of global markets that include land. Multinational corporations seeking cheap natural resources may irrevocably decimate the ecosystem and public health of a given community, but make enough profits to justify these actions to noncommunity investors, shareholders, land speculators, and governments.

Private property owners may not willingly give up property rights, even if it is for sustainability. In terms of sustainable development, land has a direct impact on the systems of nature on which future life may depend. Even countries with strong private property legal protections can have strong sustainable development initiatives, as in the case of Norway. Other countries may suffer from the impacts of climate change and rising sea levels by actually losing land they claim.

Land use law and planning, discussed later, prevents pure market forces from determining all uses of land. In the United States, and in most places in the world, environmental law is relatively new and poorly enforced. Most of the time industries are allowed to self-report emissions, citizen evidence is not allowed to challenge a given environmental permit, and urban areas are ignored. Environmental law and regulation are barely a challenge for private property systems of land tenure. Private property is not a rigid concept, but a fluid concept of law that changes over time based on the needs of the community. It is redefined, defended, altered, and confirmed in the courts. Over the years, the bundle of rights associated with private property has shifted. The contours of the right and the value of private property change to reflect the needs of local, regional, and national communities. Technological advances in our ability to affect the environment have dramatically increased the chances of committing irreversible errors that may limit rights of both future generations and current expectations of private property owners. Advances in ecosystem knowledge, increases in the technology of environmental monitoring and risk assessment, and new policies of sustainability may require that the bundle of human rights in land called private property may have to shift. Although private property is a flexible concept, it is still a human concept of law constrained by the limits of nature. It is tied to a strongly held value of individual liberty in many Western nations. As values of freedom and liberty spread with capitalism and economic development, they may confront environmental limitations imposed by local, national, and international concerns about sustainable development. Increased knowledge, increased population growth, and increased environmental impacts may make the application of the precautionary principle part of the new bundle of rights called private property.

Sustainability may present a serious challenge to private property legal systems and require that the bundle of rights again be altered. Some industries are particularly concerned with sustainability proposals because they may be currently unable to act in a sustainable manner. It may be difficult to mine for gold or oil in a fragile desert ecosystem in a sustainable manner, for example. Currently, many large private property investors concerned with the parameters of sustainable development have an emerging set of business concerns that relate to environmental considerations. First, they want to measure in capital terms the effect of sustainability on commercial real estate, especially any environmental cleanup costs. They are required to use "due diligence" to see if the sight requires cleanup, but the standards for due diligence are unclear.

Due diligence standards for sustainable development are an important environmental issue because they require accurate, timely, and verifiable environmental information about past, present, and potential ecosystem impacts. Communities and real estate developers especially want to know about sustainable designs and remediation of contaminated private property. The effects of climate change on the sustainability of a given parcel of private property are also a concern because they can affect the profit. Carbon offsets are one part of a sustainable building and construction process that are a direct response to climate change. Rising sea levels as a result of global warming are another result and can cause salt water incursion in a coastal manufacturing site or community school, for example. Rising sea levels can also cause vapor intrusion from chemically contaminated sites to migrate into the community and into sensitive ecosystems such as wetlands. Communities and real estate corporations are concerned with exactly what land stewardship means and requires because it affects the structuring of a fully integrated, sustainable property development.

PRIVATE PROPERTY IN OTHER NATIONS

Private property concepts in other nations are a complex issue tied up in politics and economics. Sometimes basic values conflict with Western capitalism. Some nations do not believe that one should lose their home to pay a loan when the home is used as collateral for a mortgage. In these countries money-lending institutions do not like to lend money for homes and the home ownership rate is much lower. The economic power of land is lower when it cannot be seized by the lender for failure to pay the lender.

In other nations, many places simply do not have an address or recorded description of the property. This makes it difficult for lending institutions to use the land as collateral for loans, often necessary for other types of economic development. Some nations have a tradition of requiring the owner of the property to take care of squatters who are on the land. In many nations the rule of law is weak, so that the same parcel of land could be sold more than once to different buyers. On top of any property rights, overlaying treaties may give various rights to indigenous peoples.

An international effort to categorize private property began in 2007. In a ground-breaking study conducted by Alexandra C. Horst, the 2007 International Property Rights Index was developed. A monumental task, this study analyzed legal property rights across the globe. It compares countries according to the strength and effectiveness of their property rights protection (www.InternationalPropertyRightsIndex.org).

Many objective and subjective factors went into this continuing assessment. These factors try to assess some dynamic characteristics, but not necessarily validate them in the field. Many nations seeking to develop may overstate their legal protections of land to increase foreign investment and the lending potential of their land mass. Private property is a dynamic bundle of rights that can be altered by law to accommodate considerations of sustainable development.

PROPERTY AS A FLEXIBLE LEGAL THEORY: ENOUGH FOR SUSTAINABILITY?

Most of the ideological and political activities of U.S. courts in land use are focused on adjusting the boundaries between public and private interests. The result of these conflicts is the laws and rules that both detract from some of the rights and privileges of private property owners and secure them from incursion from the government and others. Private property rights in the United States are continuously refined and confirmed in courts, legislatures, government agencies, business transactions, and inheritances. *See also* **Volume 1, Chapter 2: Human Values and Goals; Volume 1, Chapter 4: Private Property.**

References

Freyfogle, Eric T. 1993. *Justice and the Earth: Images for Our Planetary Survival.* New York: The Free Press.

Freyfogle, Eric T. 2003. *The Land We Share: Private Property and the Common Good.* Washington, DC: Island Press.

Fuchs, D. A. 2003. *An Institutional Basis for Environmental Stewardship: The Structure and Quality of Property Rights.* New York: Springer.

Haar, Charles M. and Michael Allen Wolf. *Land Use Planning and the Environment.* 2009. Washington, DC: Island Press.

Pozzo, Barbara, ed. 2007. *Property and Environment: Old and New Remedies to Protect Natural Resources in the European Context.* Durham, NC: Carolina Academic Press.

Raymond, Leigh Stafford. 2003. *Private Rights in Public Resources: Equity and Property Allocation in Market-Based Environmental Policy.* Washington, DC: Resources for the Future.

National Security and Confidential Business Information

Information is essential in making sustainable choices. Government has the power to assemble information on a wide array of topics to perform its duties. Information about the environment is assembled in many offices of government. Sometimes that information is publicly available. Timely public access to environmental information is essential knowledge for residents. Limits are regularly placed on the ability of the public to access information gathered by the government. One significant constraint on public access to governmental information gathered about the environment is the ability of businesses that discharge waste and pollution to shield their activities from public scrutiny to the extent that it involves confidential business information. This is information that the business asserts must remain secret to protect their trade secrets. Another significant constraint on public access to governmental information is national security. This involves information that the government asserts must remain secret to protect the nation's own health, safety, and welfare. These assertions of a right to secrets can prevent the true and full accounting for information about what is in our land, air, and water.

BUSINESS AND TRADE SECRETS

Sustainable development attracts many entrepreneurs, individuals, and businesses seeking new ways of doing business in a sustainable manner that still makes a profit. In a sense sustainable development is a new growth industry. As such new processes and products are created. Traditionally, these are protected by the legal process of patent and trade secret. Sustainable development is under particular scrutiny because if a new processes is sustainable, traditional business methods of protecting it may contravene public policies of disclosure necessary to protect the environment and the threat of irreparable damage that would impair the ability of future generations to enjoy it.

Patents and trade secret protection are meant to encourage private businesses to do the research and development necessary to move society forward. By giving special protection to the first business to market with a new good or service, public policy protects their initial profits to cover the costs of research and development. Patents of life are particularly controversial because they could have unintended impacts on the environment in ways that jeopardize future sustainability. Patents of traditional knowledge of indigenous peoples are another highly controversial area.

Patents and trade secrets are not self-operating. They require the business person to take affirmative steps to claim the product or process as their own. Trade secrets require that the information have commercial value, and that the information or how the information is used, be a secret. The potential owner of the trade secret must take reasonable steps to ensure its secrecy. This can mean restricting access to its research facilities, preventing all publication and dissemination of information related to the secret, and requiring employees who work on the secret to sign nondisclosure agreements. In the case of sustainability secrets, much of this information can be environmental. It can relate to new crops, new ways of remediating environmental areas of toxicity, new ways to construct buildings and roads, or new ways to create energy. This is a problem in sustainable development because the idea seeking trade secret protection may need to be disclosed to be useful. It simply may not be able to receive trade secret protection. This is one argument that many sustainability advocates use to promote greater governmental intervention.

The other traditional method of idea protection is the patent system. Nations differ in their approaches and levels of enforcement of patents. Under patent protection, the entrepreneur must disclose the idea, and only if that idea is new and novel does the government patent office issue the patent. Generally, an idea that is not new and is used for one year loses the ability to get a patent. If a patent is pending while the U.S. Patent Office is considering the patent application, more time is granted.

The problem of sustainable development as a public policy and private profit-making idea is problematic. This problem was the creative push for the Designers Accord. Designers face all types of environmental challenges. Not only do designers design buildings and follow LEED (Leadership in Energy and Environmental Design) design criteria to be sustainable, they design cloths, products, manufacturing processes, and art. They use plants, inks, and chemicals in processes that create waste and pollution. The Designers Accord is a set of voluntary business principles that tries to make sustainability an early part of the design process. It has been signed by more than 3,500 designers, some of the biggest design firms, and endorsed by the major design trade associations. These Accords were created by IDEO, a business design firm founded in 1991. IDEO is a design and consulting business that uses a human-centered and design-based approach to help organizations innovate and grow. Sustainability is made an early part of the process by speaking with every new client about it first, and by sharing knowledge about sustainability with others in the industry. In the firm challenges the traditional notion of trade secrets and patents and instead shares best practices based on shared experiences. Reluctant clients and lack of environmental literacy by designers have been impediments to sustainable development in the design industry. The Design Accords seek to alter this problem. The guidelines to implement the accords encourage proactive dialogues with all clients about the environmental impacts of their work and sustainable alternatives. Designers are to share knowledge with others in the industry to gain a competitive advantage, instead of the other way around.

MILITARY SECRETS

Military groups all over the world have significant environmental impacts. It is part of the nature of war to create, or threaten to create, environmental impacts that may irreparably affect systems of nature on which future life depends. Military organizations use many types of materials and processes aside from war that are used in the defense of nations. If these materials and processes are known to outside groups, then national defense is weaker. The need to keep them secret is based on national defense, even though they may have significant environmental impacts.

Regular military operations have environmental impacts. The routine deployment of troops in training and mobilization exercises regulate the use of petrochemicals. Weapons testing creates severe environmental impacts and may leave heavy metals in the land and water. Waste disposal is a large issue and often must be secret. Building, closing, and realigning military bases can have environmental impacts. For example, when an air field on a base closes, the toluene used for wing deicer often seeps into the water systems of the nearby communities.

The military is becoming more open to sustainability. In November 2008, the U.S. Army released its first annual sustainability report. The report covers the period from 2004–2007. The report describes some new developments in the area of sustainability. Sixteen Army installations have comprehensive Installation Sustainability Plans. Seventy-eight percent (301) of the Army Construction Projects in 2007 were designed to LEED new construction sustainability standards. All of the army's 161 installations have an environmental management system; 31 percent of them apply the International Standards Organization standards. They have decreased energy use by 8.4 percent. The U.S. Army still generates large amounts of waste. From 2003 to 2006, there was a 35 percent increase in hazardous waste generation and an 8 percent increase in pounds of hazardous waste generated per $1,000 net cost of army operations. Overall there was an 11 percent increase in toxic release inventory releases during the same time.

The concern for sustainability pushed the army to report and plan around sustainable development principles even in the face of a tradition of military secrets. The express purpose of the army's first report is to inform and engage communities, business, and other interested stakeholders on their status and progress in advancing the principles of sustainability in operations, installations, systems, and community engagements. The army expressly embraces the Triple Bottom Line of Sustainability by aligning its missions to environmental stewardship, community health, and economic growth. Their first report will serve as a baseline for future measures of sustainability. (For the full report go to Army Environmental Policy Institute: www. aepi.army.mil.)

Military organizations will still face environmental controversies, and the need for secrets. Over time, the environmental impacts of any secrets become known because the impacts begin to accumulate in the environment. The effects of nuclear testing in the Bikini Islands became known as attempts made decades later to make them habitable uncovered a much larger extent of ecosystem damage. The effects of weapons testing became apparent in Hawaii and Puerto Rico as the environmental impacts became known. Military technology also creates environmental impacts with unintended impacts that may threaten environments. Military technology is often one of the most closely guarded secrets. From laser-guided nuclear missiles to coal-fired jet bombers run by computers, the potential threat to sustainability becomes greater as these weapons become more powerful. The research and development, and sometimes their distribution and creation, remain secretive. Older military technology does not just go away. Although it may no longer be a secret, it becomes used in other countries. There it may also have powerful environmental impacts that can threaten sustainability. *See also* **Volume 1, Chapter 4: Information.**

Inertia: Sunk Costs, Old Technology, Political Will

Policy decisions are implemented by both action and inaction. Inaction is a choice to keep things as they are, even when the failure to act will predictably lead to a state of systemic chaos. Sometimes this tendency toward chaotic states is called entropy. Many factors in human economies can paralyze action, even when inaction probably will result in collapse or chaos. Factors contributing to such inertia include uncertainty, sunk costs in existing infrastructure that is now paid for and completed, and the political costs of change.

UNITED NATIONS

The idea that nations should come together for any purpose other than trade or war was a radical one for most of human history. Many nations preferred to be isolated and defended their borders from all encroachment. President Franklin D. Roosevelt first used the term *united nations* in the Declaration by United Nations on January 1, 1942. This declaration occurred during World War II when 26 nations agreed to continue fighting against the Axis powers.

Nations had begun to meet to try to negotiate peace before that time. The International Peace Conference was held in The Hague in 1899 in order to stop wars and develop rules for war. From this conference the Convention for the Pacific Settlement of International Disputes and the Permanent Court of Arbitration were formed. Before the United Nations was the League of Nations, which was established under the Treaty of Versailles in 1919 after World War I. The purpose of the League of Nations was to prevent wars and settle international disputes. When World War II began the League of Nations ended.

On June 26, 1945, 50 countries attending the United Nations Conference on International Organization in San Francisco met to write the United Nations Charter. After a significant number of nations signed and ratified the charter, the United Nations was created on October 24, 1945. Every year United Nations Day is celebrated on this day.

References

Fasulo, Linda. 2004. *An Insiders Guide to the UN.* New Haven, CT: Yale University Press.
United Nations. 2004. *Basic Facts about the United Nations.* New York: United Nations.

The Role of The UN

The UN is not a government in the strict sense; it is an association of governments. Together the member states explore ways in which they may act collectively on issues affecting global peace and security. Sometimes, the UN functions to assist member nations in articulating norms of behavior. When these norms are not written into enforceable treaties, this kind of activity is called soft law. Soft law can be operated on voluntarily by individual groups and by nongovernmental organizations. Hard law is an international norm of behavior backed by enforceable treaties. Sometimes soft law becomes hard law over time as international norms first articulated in soft law come to be expected and relied on.

The UN works in a number of ways to promote and develop soft and hard law around the issues of sustainable development. It sponsors scientific research and publishes information critical to policy development. It also sponsors many conferences at which soft law principles of sustainability emerge, and consensus develops. It promotes the development of nongovernmental organizations with goals compatible with these principles. In addition, it offers member governments consulting services, work plans, and training.

UN Environmental Programme (UNEP)

UNEP coordinates all the environmental activities of the UN. It began at the UN Conference on the Human Environment in Stockholm in 1972. Many scientists and environmental activists were concerned that many environmental problems spanned the boundaries of nations and that rising world populations could make these problems worse in ways no one nation could effectively handle. After the conference the UN passed resolution 2997 in December 1972, establishing the UNEP as a permanent institution charged with protection and improvement of the environment. Resolution 2997 also charges the UNEP with the missions of promoting international cooperation on the environment; reviewing global environmental issues so that governments give them adequate consideration; promoting the acquisition, assessment, and exchange of environmental knowledge; and reviewing the environmental impact of environmental policies on developing countries.

The UNEP is governed by a governing council of 58 members elected by the UN General Assembly for three-year terms. Member seats are allocated on a global regional basis. The headquarters is in Nairobi, Kenya, with six regional offices around the world. It has seven divisions: Early Warning and Assessment; Environmental Policy Implementation; Technology, Industry, and Economics; Regional Cooperation; Environmental Law and Conventions; Global Environmental Facility Coordination; and Communication and Public Information. These divisions are a big player in major international environmental initiatives. They publish many books and reports. UNEP's medium term strategy for 2010–2013 is to prioritize climate change, disasters and conflicts, ecosystem management, environmental governance, harmful substances and hazardous wastes, and resource efficiency-sustainable consumption and production. For more information on UNEP, see www.unep.org.

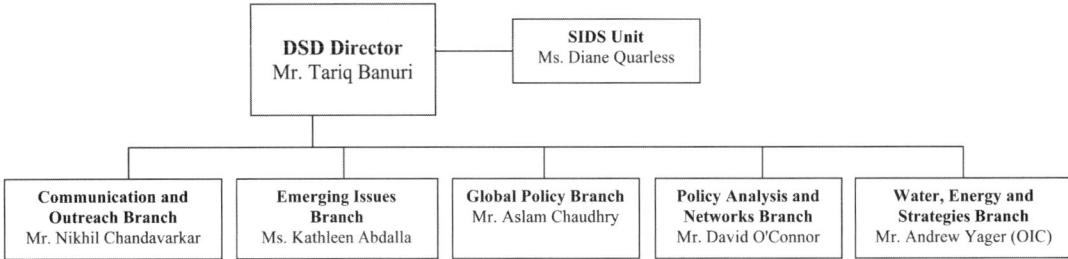

FIGURE 1.13 • UN Environment Programme Organigram.

FIGURE 1.14 • United Nations Conference on Environment and Development (UNCED) also known as the Rio Summit, Earth Summit. Opening ceremonies at the Earth Summit in Rio de Janeiro, Brazil, were held on Thursday, June 3, 1992. AP Photo/Eduardo DiBaia.

UNITED NATIONS CONFERENCES ON ENVIRONMENT AND DEVELOPMENT (UNCED)

UNCED have become the forums in which key concepts of sustainable development have been turned into "hard" international law including implementable policy statements and multilateral treaties. These agreements and statements resulting from these conferences are often identified by their host city. UNCED was renamed United Nations Environmental Programme (UNEP) in 1972.

In 1987, Gro Harlem Brundtland, then the prime minister of Norway, authored a report for the World Commission on Environment and Development called, "Our Common Future." In it, she linked the traditional goals of development to the environment and future generations. Her definition has become the most widely quoted definition of sustainable development. She wrote that sustainable development is "development that meets the needs of the present without compromising the ability of future generations to meet their own needs."

WORLD COMMISSION ON ENVIRONMENT AND DEVELOPMENT

The United Nations General Assembly created the World Commission on Environment and Development in 1983. It is an independent UN committee of 22 members and is chaired by the prime minister of Norway. Its charge is to examine the global environment and development, assess critical environmental problems, develop proposals for solving them, and raise international understanding of them. The commission laid the legal foundation for the UN Convention on Biological Diversity in 1986. The part of this convention that relates to sustainability is as follows:

Article 10. Sustainable Use of Components of Biological Diversity Each Contracting Party shall, as far as possible and as appropriate:

(a) Integrate consideration of the conservation and sustainable use of biological resources into national decision-making;

(b) Adopt measures relating to the use of biological resources to avoid or minimize adverse impacts on biological diversity;

(c) Protect and encourage customary use of biological resources in accordance with traditional cultural practices that are compatible with conservation or sustainable use requirements;

(d) Support local populations to develop and implement remedial action in degraded areas where biological diversity has been reduced; and

(e) Encourage cooperation between its governmental authorities and its private sector in developing methods for sustainable use of biological resources.

The World Environment and Development Commission is most famous for laying the groundwork for modern views of sustainability. That view was expressed in their report, "Report of the World Commission on Environment and Development: Our Common Future" issued March 20, 1987. An excerpt from that report follows:

A Call for Action

101. Over the course of this century, the relationship between the human world and the planet that sustains it has undergone a profound change.

102. When the century began, neither human numbers nor technology had the power radically to alter planetary systems. As the century closes, not only do vastly increased human numbers and their activities have that power, but major, unintended changes are occurring in the atmosphere, in soils, in waters, among plants and animals, and in the relationships among all of these. The rate of change is outstripping the ability of scientific

disciplines and our current capabilities to assess and advise. It is frustrating the attempts of political and economic institutions, which evolved in a different, more fragmented world, to adapt and cope. It deeply worries many people who are seeking ways to place those concerns on the political agendas.

103. The onus lies with no one group of nations. Countries of the global South face the obvious life-threatening challenges of desertification, deforestation, and pollution, and endure most of the poverty associated with environmental degradation. The entire human family of nations would suffer from the disappearance of rain forests in the tropics, the loss of plant and animal species, and changes in rainfall patterns. Industrial nations face the life-threatening challenges of toxic chemicals, toxic wastes, and acidification. All nations may suffer from the releases by industrialized countries of carbon dioxide and of gases that react with the ozone layer, and from any future war fought with the nuclear arsenals controlled by those nations. All nations will have a role to play in changing trends and in righting an international economic system that increases rather than decreases inequality that increases rather than decreases numbers of poor and hungry.

104. The next few decades are crucial. The time has come to break out of past patterns. Attempts to maintain social and ecological stability through old approaches to development and environmental protection will increase instability. Security must be sought through change. The Commission has noted a number of actions that must be taken to reduce risks to survival and to put future development on paths that are sustainable. Yet we are aware that such a reorientation on a continuing basis is simply beyond the reach of present decision-making structures and institutional arrangements, both national and international.

105. This Commission has been careful to base our recommendations on the realities of present institutions, on what can and must be accomplished today. But to keep options open for future generations, the present generation must begin now, and begin together.

106. To achieve the needed changes, we believe that an active follow-up of this report is imperative. It is with this in mind that we call for the UN General Assembly, upon due consideration, to transform this report into a UN Programme on Sustainable Development. Special follow-up conferences could be initiated at the regional level. Within an appropriate period after the presentation of this report to the General Assembly, an international conference could be convened to

review progress made, and to promote follow up arrangements that will be needed to set benchmarks and to maintain human progress.

107. First and foremost, this Commission has been concerned with people—of all countries and all walks of life. And it is to people that we address our report. The changes in human attitudes that we call for depend on a vast campaign of education, debate, and public participation. This campaign must start now if sustainable human progress is to be achieved.

108. The Members of the World Commission on Environment and Development came from 21 very different nations. In our discussions, we disagreed often on details and priorities. But despite our widely differing backgrounds and varying national and international responsibilities, we were able to agree to the lines along which change must be drawn.

109. We are unanimous in our conviction that the security, well-being, and very survival of the planet depend on such changes, now.

(The full text of this monumental report is available at www.un.org/documents/ga/res/42/ares42–187.htm.)

UN FRAMEWORK CONVENTION ON CLIMATE CHANGE

The Convention on Climate change has been ratified by 192 nations and went into effect on March 21, 1994. It sets forth an overall framework for an approach to climate change policies. It does not require emission limitations but does encourage them. It implicitly recognizes the complexity of the climate change problem in terms of its many causes and the many nations and their governments and their approach to it. By agreeing to this convention, nations are expected to collect and share information on greenhouse gas emissions, national climate change policies, and best environmental practices for handling greenhouse gas emissions. Ratifying nations are expected to develop new national strategies for controlling greenhouse gas emissions and adapting to climate changes. Part of this expectation is to provide financial and technological support to developing countries to combat the causes and effects of climate change. Ratifying nations are expected to cooperate with one another and with developing nations in preparing for the impacts of climate change.

U.S. INVOLVEMENT WITH THE UNITED NATIONS FRAMEWORK CONVENTION ON CLIMATE CHANGE

The United States signed the convention in June 1992, and ratified it that year in October. The United States submitted the required climate change reports in 1994, 1997, 2002, and 2007. The 2007 U.S. Climate

Action Report gave an inventory of U.S. greenhouse gas emissions, an estimate of mitigation policies on future emissions, and a description of U.S. involvement in international programs. The fourth report discusses the U.S. Climate Change Science Program, the U.S. Climate Change Technology Program, and the U.S. Integrated Earth Observation System.

The major controversy with the United Nations Framework Convention on Climate Change is that it was modified by the Kyoto Protocol. The Kyoto Protocol was adopted in Kyoto, Japan on December 11, 1997 and began working for the 184 nations that ratified it on February 16, 1997. The United States has signed the protocol but has not ratified it. The protocol is more stringent in that it has stronger mitigation and reporting measures that are more legally binding. It sets binding emissions limits on greenhouse gases for 37 industrialized nations and the European community. These reductions aim at 5 percent of 1990 levels of emissions to be attained from 2008 to 2012.

The Kyoto Protocol uses three market-based mechanisms to try to decrease greenhouse gas emissions: emissions trading, clean development mechanisms, and joint implementation. Furthermore, actual emissions must be monitored under the protocol with registry systems, international transaction logs, and thorough reporting. Ratifying nations are to help other nations adapt to climate changes caused by global warming and climate change.

The first mechanism, making a commodity of greenhouse gas reductions, is a strong commitment under the Kyoto Protocol. These reductions are called "assigned amounts" and are the amounts of greenhouse gas allowed during the 2008–2012 period. Countries that do not use all their emissions are allowed to trade their assigned amount units they do not use. The major greenhouse gas emitted under the Kyoto Protocol is carbon dioxide. It is the major greenhouse gas that is traded, and these international emissions trading programs are called carbon markets. Other trading mechanisms under the Kyoto Protocol equate to a unit of carbon dioxide removed from emissions inventories. These are called "removal units." They are based on land use, land use change, and forestry changes that decrease the amount of carbon dioxide emitted as a greenhouse gas. These activities that reduce the greenhouse gas emission into the air through an increase in plant processes on land are called "sinks." These have been calculated on a global scale. Forests are large sinks for carbon dioxide and store about 283 Gt of carbon in its living structure, 38 Gt in the dead wood, and about 317 Gt in its topsoil. In 2005, the global carbon sink in forests was estimated to be about 638 Gt. Since this is more than the carbon in the atmosphere, carbon sequestration by forests is a significant factor controlled by human activities. This is why deforestation is a crucial issue in global warming. If the trees are harvested and burned, the amount of carbon sequestered in the soil and vegetation is much less and the amount in the atmosphere is much more. This also highlights the importance of U.S. ratification of the Kyoto Protocol, as the U.S. is one of the largest consumers of wood products and nonrenewable energy sources such as oil.

The second market-based mechanism under the Kyoto Protocol for controlling greenhouse gas emissions is called a clean development mechanism. The protocol authorizes a developed country to work with a developing country to reduce greenhouse gases by letting the country with an emission limitation earn more rights to emissions by reducing the emissions of the developing country. This is a strong push for sustainable development because many developing countries have weak power sources and, with help from developing countries, can economically develop in ways that do not emit greenhouse gases. These projects earn certified emission reduction credits that can be bought and sold. This is the first internationally standardized emissions trading scheme. Projects that use this scheme are called clean development mechanisms. They are generally projects that go beyond what would have otherwise occurred. There are rigorous registration and approval requirements for these projects. More than 1,000 projects have been approved, saving 2.7 billion tons of carbon dioxide equivalents since 2006.

There are still other issues than the lack of U.S. ratification of the Kyoto Protocol. There is an issue of international transportation emissions from ships and airplanes that are not currently counted in national greenhouse gas registries. Many of the issues relate to accurate assessment of environmental emissions and activities. Many countries do not fully evaluate their environmental emissions or activities in order to court economic development. Short-term profits are greater when environmental impacts are ignored. There is a great temptation by developing nations to ignore these impacts with the sale of natural resources or use of nonrenewable energy. If all the developing nations emitted as much as the developed nations have to date, the impact on global warming and climate change would be much greater and the threat to sustainability would increase. The Kyoto Protocol seeks to have developed nations help developing nations achieve economic development without increasing greenhouse gas emissions.

The Kyoto Protocol is considered the main international agreement and method for affecting a reduction of human-caused greenhouse gas emissions. At the end of 2012, the Kyoto Protocol will be revisited. The pressure will be on the United States to ratify it because the Unites States is one of the largest greenhouse gas emitters and because it has benefited from the development that caused global warming and climate change. The climate change goals in the Intergovernmental Panel on Climate Change called for strict emissions limitations and may be included in the 2012 reiteration of the Kyoto Protocol. (For the Kyoto Protocol, see unfccc.int/resource/docs/convkp/kpeng.html.)

Intergovernmental Panel on Climate Change (IPCC)

The IPCC began when the UNEP and the World Meteorological Organization met in 1988 to review the entire scientific and technical peer-reviewed literature on global climate change. The IPCC does not perform any original scientific research. Member countries choose the scientists who participate. These scientists are considered the leading experts on climate change and their reports are relied on heavily in all world pronouncements on the issue. Many nations base their environmental policies and programs on their reports. Most reports are accessible to policymakers. The reports generally include discussion of future development scenarios relating to social, economic, technological, and energy use. They specifically include analyses of future greenhouse gas emissions.

Currently, the IPCC has three working groups and the Task Force on National Greenhouse Gas Inventories. The mission of Working Group One is to study the research on the physical science basis of climate change. The mission of Working Group Two is to handle the research on climate change impact, adaptation, and vulnerability. Working Group Three handles mitigation of climate change. All groups face powerful methodological and data challenges.

The IPCC has been criticized for being too cautious in its scientific approach to climate change. Part of the scientific methods is to count only those changes that fall within prescribed scientific levels. Some ICC reports limit projections of temperature changes to those that fall within a 90 percent confidence level. This discounted a small probability of larger scale unpredictable changes, such as methane gas releases from the Arctic tundra or complete loss of the Antarctica and Greenland ice fields. The loss of the reflective ice (called the albedo effect) and the absorption of the sun's heat by the then exposed water could dramatically increase global warming. In terms of sustainability, some of the scientific assumptions of the IPCC could irreparably harm the systems of life on which future generations depend. They assume that benefits for future generations should be discounted relative to the costs on present generations. They also make little consideration for the impacts on the most vulnerable nations, either in terms of the small probability of nonlinear, rapid environmental changes or in the impacts of an intergenerational cost-benefit analysis. The last set of criticisms is based on the fact that the scientific literature is technical in nature and is peer reviewed. It is not from people who may be experiencing the environmental changes first hand, as is the case with many indigenous peoples. Much of the research comes from studies performed by those with a stake in the profit made from a given natural resource, such as timber, minerals, and water. Further, the nature of peer review limits the publication of a given work to a traditional paradigm, or way of thinking. Peers are often those of a singular discipline, so that a given piece of environmental or ecological research that is outside of the usual disciplinary boundaries will be rejected. In this way, the IPCC is considered by some to be self-limiting, and may be underpredicting the potential for climate changes. In response to some of these criticisms, the IPCC issued a decision framework prioritizing assessment reports and other measures to protect the credibility of the organization in 2008.

Al Gore

Al Gore was born in 1948 into a political family. His father, Albert Gore, served in the U.S. Congress (both houses) from 1939 until 1971. His family owned a tobacco farm in Tennessee where Gore worked in the summers. He attended a private school in Washington DC before attending Harvard University, where he earned a degree in government with honors. After serving in Vietnam as a combat journalist, he attended Vanderbilt University School of Divinity and later Vanderbilt Law School. He earned no further degrees because he decided to run for his father's former seat in the House of Representatives. He served there for three terms, and later won a seat in the U.S. Senate.

His political endeavors did not end there, however. From 1992–2000, Gore served as vice president. It was during this time that Gore became known for his commitment to the environment and to stopping global warming. During his stint in office, Gore was involved in the Kyoto Protocol negotiations and disarmament, and he initiated a think-tank to review the health of the world's ecosystems. In 2000, Gore ran for president against George W. Bush, winning the popular vote, but losing the electoral vote in a lawsuit affected by the U.S. Supreme Court's decision to stop counting votes.

Subsequently, Gore's film on climate change, *An Inconvenient Truth*, won an Academy Award for Best Documentary. He also shared the 2007 Nobel Peace Prize with the Intergovernmental Panel on Climate Change "for their efforts to build up and disseminate greater knowledge about man-made climate change, and to lay the foundations for the measures that are needed to counteract such change."

References

Gore, Albert. 1992. *Earth in the Balance: Ecology and the Human Spirit.* Boston: Houghton Mifflin.

Gore, Albert. 2006. *An Inconvenient Truth: The Planetary Emergency of Global Warming and What We Can Do about It.* Emmaus, PA: Rodale Press.

Gore, Albert. 2007. *The Assault on Reason.* New York: Penguin Press.

FIGURE 1.9 • Nobel Peace Prize winners Al Gore, center, and Chairman of the Intergovernmental Panel on Climate Change Dr. Rajendra K. Pachauri, right, receive their medal and diploma from Nobel Committee Chairman Ole Danbolt Mjoes, left, at City Hall in Oslo, December 10, 2007. AP Photo/John McConnico.

The "Earth Summit," the Rio Declaration, and Agenda 21

The United Nations hosted a Conference on Environment and Development in Rio de Janiero, Brazil, in 1992. The conference itself is often referred to as the Earth Summit. In a declaration called the "Rio Declaration," member nations including the United States, agreed to link future development to sustainable development principles. This conference produced a work plan for sustainable development that has guided all subsequent efforts to implement sustainability and sustainable development. This work plan is called Agenda 21 anticipating the achievements of sustainable development in the 21st century. *See also* **Volume 1, Appendix B: Rio Declaration on Environment and Development.**

AGENDA 21

Agenda 21 is a global plan of action for sustainable development. It was drafted at the 1992 United Nations Conference on Environment and Development, or Earth Summit in Rio de Janeiro, Brazil. More than 130 heads of state and more than 15,000 member of nongovernmental organizations attended the meeting. The result of the conference was Agenda 21 and the Rio Declaration on Environment and Development (Rio Declaration). The two nonbinding plans dovetail one another and therefore should be examined together for a clearer picture of the mission they hope to accomplish. The Rio Declaration is a statement of 27 principles for sustainable development; Agenda 21 is a longer 40-chapter plan of action to conform to the principles in the Rio Declaration. The mission of the Earth Summit was to synthesize and integrate environmental and development issues. The idea was to incorporate environmental protection into the global understanding of what is required for progress.

The principles of the Rio Declaration are:

- Principle 1: Human beings are at the centre of concerns for sustainable development. They are entitled to a healthy and productive life in harmony with nature.

- Principle 2: States have, in accordance with the Charter of the United Nations and the principles of international law, the sovereign right to exploit their own resources pursuant to their own environmental and developmental policies, and the responsibility to ensure that activities within their jurisdiction or control do not cause damage to the environment of other States or of areas beyond the limits of national jurisdiction.

- Principle 3: The right to development must be fulfilled so as to equitably meet developmental and environmental needs of present and future generations.

- Principle 4: In order to achieve sustainable development, environmental protection shall constitute an integral part of the development process and cannot be considered in isolation from it.

- Principle 5: All States and all people shall cooperate in the essential task of eradicating poverty as an indispensable requirement for sustainable development, in order to decrease the disparities in standards of living and better meet the needs of the majority of the people of the world.

- Principle 6: The special situation and needs of developing countries, particularly the least developed and those most environmentally vulnerable, shall be given special priority. International actions in the field of environment and development should also address the interests and needs of all countries.

- Principle 7: States shall cooperate in a spirit of global partnership to conserve, protect and restore the health and integrity of the Earth's ecosystem. In view of the different contributions to global environmental degradation, States have common but differentiated responsibilities. The developed countries acknowledge the responsibility that they bear in the international pursuit to sustainable development in view of the pressures their societies place on the global environment and of the technologies and financial resources they command.

- Principle 8: To achieve sustainable development and a higher quality of life for all people, States should reduce and eliminate unsustainable patterns of production and consumption and promote appropriate demographic policies.

- Principle 9: States should cooperate to strengthen endogenous capacity-building for sustainable development by improving scientific understanding through exchanges of scientific and technological knowledge, and by enhancing the development, adaptation, diffusion and transfer of technologies, including new and innovative technologies.

- Principle 10: Environmental issues are best handled with participation of all concerned citizens, at the relevant level. At the national level, each individual shall have appropriate access to information concerning the environment that is held by public authorities, including information on hazardous materials and activities in their communities, and the opportunity to participate in decision-making processes. States shall facilitate and encourage public awareness and participation by making information widely available. Effective access to judicial and administrative proceedings, including redress and remedy, shall be provided.

- Principle 11: States shall enact effective environmental legislation. Environmental standards, management objectives and priorities should reflect the environmental and development context to which they apply. Standards applied by some countries may be inappropriate and of unwarranted economic and social cost to other countries, in particular developing countries.

- Principle 12: States should cooperate to promote a supportive and open international economic system that would lead to economic growth and sustainable development in all countries, to better address the problems of environmental degradation. Trade policy measures for environmental purposes should not constitute a means of arbitrary or unjustifiable discrimination or a disguised restriction on international trade.

- Unilateral actions to deal with environmental challenges outside the jurisdiction of the importing country should be avoided. Environmental measures addressing transboundary or global environmental problems should, as far as possible, be based on an international consensus.

- Principle 13: States shall develop national law regarding liability and compensation for the victims of pollution and other environmental damage. States shall also cooperate in an expeditious and more determined manner to develop further international law regarding liability and compensation for adverse effects of environmental damage caused by activities within their jurisdiction or control to areas beyond their jurisdiction.

- Principle 14: States should effectively cooperate to discourage or prevent the relocation and transfer to other States of any activities and substances that cause severe environmental degradation or are found to be harmful to human health.

- Principle 15: In order to protect the environment, the precautionary approach shall be widely applied by States according to their capabilities. Where there are threats of serious or irreversible damage, lack of full scientific certainty shall not be used as a reason for postponing cost-effective measures to prevent environmental degradation.

- Principle 16: National authorities should endeavour to promote the internalization of environmental costs and the use of economic instruments, taking into account the approach that the polluter should, in principle, bear the cost of pollution, with due regard to the public interest and without distorting international trade and investment.

- Principle 17: Environmental impact assessment, as a national instrument, shall be undertaken for proposed activities that are likely to have a significant adverse impact on the environment and are subject to a decision of a competent national authority.

- Principle 18: States shall immediately notify other States of any natural disasters or other emergencies that are likely to produce sudden harmful effects on the environment of those States. Every effort shall be made by the international community to help States so afflicted.

- Principle 19: States shall provide prior and timely notification and relevant information to potentially affected States on activities that may have a significant adverse transboundary environmental effect and shall consult with those States at an early stage and in good faith.

- Principle 20: Women have a vital role in environmental management and development. Their full participation is therefore essential to achieve sustainable development.

- Principle 21: The creativity, ideals and courage of the youth of the world should be mobilized to forge a global partnership in order to achieve sustainable development and ensure a better future for all.

- Principle 22: Indigenous people and their communities and other local communities have a vital role in environmental management and development because of their knowledge and traditional practices. States should recognize and duly support their identity, culture and interests and enable their effective participation in the achievement of sustainable development.

- Principle 23: The environment and natural resources of people under oppression, domination and occupation shall be protected.

- Principle 24: Warfare is inherently destructive of sustainable development. States shall therefore respect international law providing protection for the environment in times of armed conflict and cooperate in its further development, as necessary.

- Principle 25: Peace, development and environmental protection are interdependent and indivisible.

- Principle 26: States shall resolve all their environmental disputes peacefully and by appropriate means in accordance with the Charter of the United Nations.

- Principle 27: States and people shall cooperate in good faith and in a spirit of partnership in the fulfillment of the principles embodied in this Declaration and in the further development of international law in the field of sustainable development.

In contrast to the Rio Principles, Agenda 21 is significantly longer and details a context-specific meaning for sustainable development.

Agenda 21 provides a starting point towards the goals of sustainable development and is based on the 27 principles described here. The primary responsibility of implementing the goals of these two documents rests with national governments; they must be willing to lead and facilitate sustainable development practices by their citizens.

Agenda 21 adopted the core principle that the major cause of environmental degradation is unsustainable process of manufacturing, and unsustainable levels of consumption mainly in industrialized nations. Agenda 21 is composed of 27 principle chapters. These remain the foundation of many governmental and nongovernmental efforts to implement sustainability at the local, national, and international level. The 27 chapters identify seven key concepts for laws and public policies to achieve sustainability:

- *Integrated Decision Making:* to consider environment, economics, and equity at the same time and of equal importance in all development decisions.

- *Sustainable Consumption and Production:* Methods of production and products should be produced without waste or pollution.

- *Intergenerational Equity:* Development decisions should not compromise the ability of future generations to meet their own needs.

- *Precautionary Principle:* When harm to the environment or human health will result from development, reasonable measures should be taken to prevent harm even if the scientific evidence is inconclusive.

- *Polluter Pays:* The costs of pollution should be paid by the source of that pollution. Governments should make sure that these costs are not paid by others by implementing fines, enforcement, and taxes.

- *Public Participation:* The best development decisions are made by people who have to live with their consequences. Governments must include these people in the decisions that affect them by direct access to the decision-making processes. ***See also:* Volume 3, Equity and Fairness, Aarhus Convention; Environmental Justice.**

- *Differentiated Leadership:* Sustainability can mean very different things depending on a country's economic and social conditions. In some countries, the need to develop is driven by the basic human needs, and in others consumption of resources far exceeds basic human needs. Sustainability may mean reduction in consumption for some, and increasing development without unsustainable methods of production in others. Leadership toward sustainability must mutually recognize different approaches for different conditions.

The precautionary principle is a moral and political principle designed to prevent serious and irreversible harm to the public or the environment. The principle aims to provide guidance for protection in the face of uncertain risk. Use of the term *precaution* indicates that a preemptive responsibility on the part of the proponent of the proposed activity to establish that their actions will not, or are unlikely to, result in significant harm. Furthermore, where there is a lack of scientific consensus that harm would not ensue, the burden of proof falls on those who would advocate taking the action. The 1998 Wingspread Statement on the Precautionary Principle summarizes the principle this way: "When an activity raises threats of harm to human health or the environment, precautionary measures should be taken even if some cause-and-effect relationships are not fully established scientifically." (The Wingspread Conference on the Precautionary Principle was convened by the Science and Environmental Health Network.)

The February 2, 2000 European Commission Communication on the Precautionary Principle notes: "The precautionary principle applies where scientific evidence is insufficient, inconclusive or uncertain and preliminary scientific evaluation indicates that there are reasonable grounds for concern that the potentially dangerous effects on the environment, human, animal or plant health may be inconsistent with the high level of protection chosen by the EU."

Reference

Dernbach, John, ed. 1992. *Stumbling Toward Sustainability*. Washington, DC: Environmental Law Institute, 45–63.

The Kyoto Protocol

Another famous UNCED conference was held in Kyoto, Japan, in 1997, resulting in the Kyoto Protocol on climate change and the limits on emissions of greenhouse gases. A majority of the nations at the conference have ratified the protocol. Although a signatory to the protocol, the United States has not moved forward with their ratification even though it is the world's largest producer of carbon dioxide, because it disagrees with the exemptions given to developing economies like China and India.

The Kyoto Protocol is the leading international agreement for affecting human-caused climate change. Most of the other international climate change studies and policies feed into the Kyoto Protocol. These protocols require honest and accurate assessments and policies about greenhouse gas emissions. They require strict limits on what developed countries are allowed to emit, and require developed countries to assist developing countries in ways that limit greenhouse gas emissions. The Kyoto Protocol are the first significant international step to meaningfully affect global natural systems. As such, their future impact on sustainability is immense. It is a new and dynamic set of principles and agreements, and will be revisited many times in the future.

The Millennium Ecosystem Assessment (MA)

The Millennium Ecosystem Assessment (MA) is, to date, the most comprehensive survey of the ecological state of the planet. The MA was called for by the then-secretary general of the United Nations, Kofi Annan, in 2000 and began its work in 2001. The MA was undertaken by an international network of scientists and other experts. More than 1,300 authors from 95 countries were involved in the MA and were divided into four working groups. Three of the working groups, Condition & Trends, Scenarios, and Responses, focused on global assessment goals whereas the fourth working group focused on subglobal assessments. The resulting assessment was divided into four technical volumes that were reviewed by experts and governments, and more than 600 individual reviewers worldwide provided approximately 18,000 individual comments. The assessment lasted four years and concluded in 2005 with its report on the consequences of ecosystem change for human well-being and the scientific basis for actions needed to enhance the conservation and sustainable uses of the Earth's resources. The MA reported four main findings:

1. Over the past 50 years, humans have changed ecosystems more rapidly and extensively than in any comparable period of time in human history, largely to meet rapidly growing demands for food, fresh water, timber, fiber, and fuel. This has resulted in a substantial and largely irreversible loss in the diversity of life on Earth.

2. The changes that have been made to ecosystems have contributed to substantial net gains in human well-being and economic development, but these gains have been achieved at growing costs in the form of the degradation of many ecosystem services, increased risks of nonlinear changes, and the exacerbation of poverty for some groups of people. These problems, unless addressed, will substantially diminish the benefits that future generations obtain from ecosystems.

3. The degradation of ecosystem services could grow significantly worse during the first half of this century and is a barrier to achieving the Millennium Development Goals.

4. The challenge of reversing the degradation of ecosystem while meeting increasing demands for services can be partially met under some scenarios considered by the MA, but will involve significant changes in policies, institutions, and practices that are not currently underway. Many options exist to conserve or enhance specific ecosystem services in ways that reduce negative tradeoffs or that provide positive synergies with other ecosystem services.

The findings basically outlined the negative impact of human actions on the Earth's natural capital. The MA showed that human's cannot take the Earth's ability to sustain future generations for granted at this rate of environmental degradation; however, if appropriate substantial action were to be taken it would be possible to reverse some degradation over the next 50 years. The MA is an assessment of data that was already available; it is unique in that it was a global assessment that presented a consensus view of the current state of the planet. Consensus is one of the cornerstones of change. Furthermore the MA also identified a number of "emergent" findings or conclusions that were the result of examining a large amount of information together.

One of the simultaneous strengths and weaknesses of the MA is the gaps in knowledge about the status of the planet's ecosystem. There is relatively limited information available at a local and national level about the status of ecosystem services and even less information about the value of nonmarketable services. There is also limited information about the economic costs of ecosystem degradation. In an increasingly market-based world, it is imperative for economies to have information and assessment of economic loss of human action. The MA created awareness of these gaps in knowledge. It is hope that it has stimulated data gathering and assessment to eliminate the gaps.

The MA work groups expect that there will be significant adoption of the MA conceptual framework that will continue to help meet the planet's sustainability needs and reverse degradation. The MA indicated that changes be instituted firmly and quickly. It was recognized that, as humanity has the power and ability to prevent the damages to the

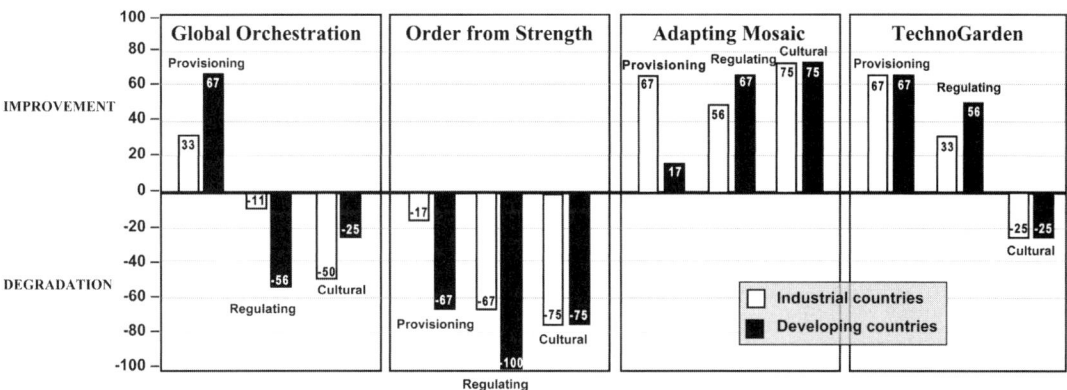

FIGURE 1.16 • Number of ecosystem services enhanced or degraded by 2050. 100 percent degradation means that all the services in the category were degraded in 2050 compared with 2000, while 50 percent improvement could mean that three of six services were enhanced and the rest were unchanged or that four of six were enhanced and one was degraded. The total number of services evaluated for each category was six provisioning services, nine regulating services, and five cultural services. Philippe Rekacewicz, Emmanuelle Bournay, UNEP/GRID-Arendal, http://maps.grida.no/go/graphic/number-of-ecosystem-services-enhanced-or-degraded-by-2050.

planet, it is also our duty to do so. One of the most important issues brought up was the effects of environmental damage to the underdeveloped and poor people of the world. The report urged the nations of the world to work harder to achieve a sustainable future. *See also* **Appendix D: The Millennium Ecosystem Assessment.**

UNITED STATES

President's Council on Sustainable Development (PCSD)

The PCSD was created in 1993 by President Bill Clinton. It was asked to find ways to "to bring people together to meet the needs of the present without jeopardizing the future." The idea mirrors the ideals of the 1987 Brundtland Commission statement on Sustainable Development. The PCSD's vision statement proclaims that, "We [the United States] are committed to the achievement of a dignified, peaceful and equitable existence." The PCSD also includes 16 "We Believe Statements" that outline the shared beliefs of the Council about the core principles of sustainable development. These beliefs include encouraging jobs, productivity, fair wages, savings, profit, information, knowledge, and education while disavowing pollution, waste, and poverty. They further encourage that change is necessary and progress must be made in an innovative and efficient way. In addition, the protection of the environment is at the forefront of sustainable development, and this can be accomplished only through collaborative decisions, increased regulations, and each individual becoming a steward of the environment on a personal and global level. The beliefs incorporate a need to link economic growth, environmental protection, and social equity and build strong communities. The basic principles of the PCSD incorporate the precautionary principle and links national and global security and prosperity to a growing economy and a healthy environment. The report issued by the PCSD is divided into seven chapters preceded by the Preface, Definition and Vision Statement, and We Believe Statement.

The following chapter summaries are all taken from the February 1996 report of the PCSD: *Sustainable America—A New Consensus for Prosperity, Opportunity and a Healthy Environment for the Future:*

Chapter 1: National Goals Toward Sustainable Development

A common set of goals that emerged from the Council's vision; the goals are truly interdependent and flow from the understanding that it is essential to seek economic prosperity, environmental protection and social equity together. These goals express in concrete terms the elements of sustainability. Alongside the goals are suggested indicators that can be used to help measure progress towards achieving them.

Chapter 2: Building a New Framework for A New Century

Future progress requires that the United States broaden its commitment to environmental protection to embrace the essential components of sustainable development: environmental health, economic prosperity and social equity and well-being. This means reforming the current system of environmental management and building a new and efficient framework based on performance, flexibility linked to accountability, extended product responsibility, tax and subsidy reform and market incentives.

Chapter 3: Information and Education

Information and education, in both formal and informal spheres, have a tremendous potential for increasing citizen awareness and ability to engage in decisions affecting their lives. Key to the strategy is managing information better, expanding access to the decision process, measuring progress towards societal goals more comprehensively and incorporating accounting measures that educate and enable decision-makers and individuals to make decisions that are more economically, environmentally, and socially sustainable. Additionally, the country's formal education system must be reformed to address sustainability better and informal education forums and mechanisms tapped to promote opportunities for learning about sustainability.

Chapter 4: Strengthening Communities

Creating a better future depends, in part, on the knowledge and involvement of citizens and on a decision-making process that embraces and encourages differing perspectives of those affected by governmental policy. Steps toward a more sustainable future include developing community-driven strategic planning and collaborative regional planning; improving community and building design; decreasing sprawl; and creating strong, diversified local economies while increasing jobs and other economic opportunities.

Chapter 5: Natural Resources Stewardship

Stewardship is an essential concept that helps to define appropriate human interaction with the natural world. An ethic of stewardship builds on collaborative approaches; ecosystem integrity; and incentives in such areas as agricultural resource management, sustainable forestry, fisheries, restoration and biodiversity conservation.

Chapter 6: U.S. Population and Sustainability

Population growth, especially when coupled with current consumption patterns, affects sustainability. A sustainable United States is one where all Americans have access to family planning and reproductive health services, women enjoy increased opportunities for education and employment, and responsible immigration policies are fairly implemented and enforced.

Chapter 7: International Leadership

The United States has both reason and responsibility to develop and carry out global policies that support sustainable development. Because of its history and power, the United States is inevitably a leader and needs to be an active participant in cooperative international efforts to encourage democracy, support scientific research, and enhance economic development that preserves the environment and protects human health.

National Environmental Policy Act (NEPA)

The express legislative purpose of NEPA is:

To declare a national policy which will encourage productive and enjoyable harmony between man and his environment; to promote efforts which will prevent or eliminate damage to the environment and biosphere and stimulate the health and welfare of man; to enrich the understanding of ecological systems and natural resources important to the Nation; and to establish a Council on Environmental Quality. NEPA section 2.

The underlying purposes of NEPA are much in line with current principles of sustainability. They are to:

1. Fulfill the responsibilities of each GENERATION as trustee of the environment for succeeding generations;

2. Assure for all Americans safe, healthful, productive, and esthetically and culturally pleasing surroundings,

3. Attain the widest range of beneficial uses of the environment with degradation, risk to health or safety, or other undesirable and unintended consequence;

4. Preserve important historic, cultural, and natural aspects of our national heritage and maintain, wherever possible, an environment which supports diversity, and variety of individual choice;

5. Achieve a balance between population and resource use which will permit high standards of living and a wide sharing of life's amenities, and

6. Enhance the quality of renewable resources and approach maximum attainable recycling of depletable resources.

NEPA, section 101(b).

The National Environmental Policy Act of 1970 puts forth the most progressive ideas about environmental protection at the time. It is not a National Sustainability Act.

NEPA has fallen far short of its lofty goals. Some advancement did occur, such as a reduction of air pollutants and the resulting decrease in acid rain in the northeastern United States and southwestern Canada. This proved to many disbelievers that a national environmental policy could really effect meaningful change in natural systems. Most courts interpret the lofty and well-meaning goals of NEPA to be aspirational. That is, they are only aspirations, not legislatively mandated goals. Some studies of court cases of NEPA have uncovered a bias in the courts cases by political party. Republican-appointed judges do not find NEPA violations and Democratic-appointed federal judges do. Currently, NEPA is riddled with "Categorical Exclusions," which are categories of actions that do have a significant impact on the environment, but that special exceptions are made by Congress or the EPA so that they are not required to even assess the environmental impacts. In the state of Oregon, 92 percent of significant environmental impacts are categorically excluded from any requirement of environmental impact assessment. This list of special exceptions only gets longer, while the environmental consequences continue to mount. State versions of NEPA, called State

What Would a National Sustainability Act Look Like?

Ideas are the most powerful things in the world. It was an idea that created the atomic bomb. It was an idea that created the National Environmental Policy Act (NEPA). And it will be an idea that creates a National Sustainability Act. Ideas often begin with hard questions. The question is, what would a National Sustainability Act really look like?

A National Sustainability Act would probably begin with the best of what there is presently in the NEPA. The laudatory goals of NEPA that courts now describe as "aspirational" would need to be implemented. That would require an increased priority in monitoring past, present, and future impacts on the environment. It would also require that when decisions must be made about anything with an environmental impact, any risk that the knowledge of these impacts is inaccurate or incomplete would delay a decision. The concept of environment would need to be expanded to include all ecological systems that define life on earth. It would probably require a decreased priority on social goals that gave us a high quality of life, such as economic systems that value capital and political power over presently unknown environmental risks. It will require humans to protect from extinction all known species that currently exist on Earth.

In other words, such a legislative act may require changes that would currently be perceived as radical, unworkable, expensive, and impractical. These were the same criticisms of NEPA when first enacted, and it may that the short-sighted and self-interested vision of humans that continue to degrade natural systems that all future life depends on will continue. But as knowledge, population, and waste march forward, new ideas emerge about how to transition into a sustainable future, and an idea is a powerful thing.

Environmental Protection Acts or SEPAs have less coverage of significant environmental impacts and about the same scale of categorical exclusions. In the rare case that a full environmental impact assessment is required, manipulations occur with regard to the actual physical area to be studied, called the study area. In one case in Eugene, Oregon, where a new federal courthouse was built, the study area was changed to exclude the underground storage tanks of a gas station formerly on the site. Many old underground storage tanks for gasoline and other petrochemical products leak. The courthouse is situated near the Willamette River, and the leaking material from such a tank could go to the river. If it had been included in the study area, the environmental impacts would have been greater. They may have required the federal agency responsible for building the courthouse to mitigate, or clean up, any leaks to the river. This would have cost more money and brought unwelcome public attention to the project. It may have also resulted in a cleaner river with less environmental impact on endangered species. This is one example of a dynamic under NEPA that unfolds on a daily basis.

Even the strongest component of NEPA, environmental impact statements, is only guidance to the business decision. The EPA administrator does not have to follow the recommendations of the environmental impact statement. It does bring substantial social resources to the environmental consequences of a small number of decisions.

Environmental Assessments (Environmental Impact Statements)

Most of sustainability policy hinges on environmental assessments of the human impacts on the environment. It has been assessments of the environment that warn of the degradation of natural systems. Environmental assessment policies determine if, when, and what kind of environmental mitigation is necessary. Most environmental assessments relied on by governments are based on some degree of science. There are international standards of environmental impact assessment such as those put forward by the United Nations Social, Educational, Social, and Cultural Organization.

On June 27, 1985, the European Council of Ministers adopted a rule that required its members to adopt environmental assessment procedures. Its members include Austria, Belgium, Denmark, Finland, France, Germany, Greece, Ireland, Italy, Luxembourg, The Netherlands, Portugal, Spain, Sweden, and the United Kingdom. They include a threshold determination of whether the environmental impacts are significant, and are applicable to both government and private projects. They have a list of projects that require an environmental impact statement (EIS) and another list of projects for which an EIS is discretionary.

Communities do their own environmental assessments, and citizen monitoring of the environment is on the increase. Industries also do their own environmental assessments. There are many kinds of environmental assessments. Models of environmental assessment that ignore, diminish, or underreport actual environmental impacts are inadequate for sustainability purposes. It is likely that many of the currently used environmental assessment models will be the springboard for sustainability environmental assessment models.

THE U.S. MODEL

The first U.S. federal law requiring environmental impact assessments was the National Environmental Policy Act of 1970 (NEPA). Many states have since developed their own State Environmental Policy Acts, and some tribes have developed their own Tribal Environmental Policy Acts. An overall critique of the environmental assessment policy context is that it does not cover enough of those human activities that result in environmental impacts. For example, they often fail to measure cumulative impacts or ecosystem impacts over long periods. Some environmentalists want all environmental impacts covered by new policies such as a Municipal Environmental Impact Statement. Pushing the NEPA environmental assessment model to cover all environmental impacts that threaten natural systems on which all life depends would enrich it to sustainability levels.

NEPA requires a detailed environmental impact statement for major federal actions significantly affecting the quality of the human environment.

U.S. ENVIRONMENTAL IMPACT ASSESSMENTS ARE ADVISORY

A fundamental weakness of NEPA is that the environmental assessment is not a document that determines policy. The environmental assessment is ultimately advisory only. This would make it unsuitable for purposes of sustainability because it would not be action oriented enough. Even so, much of the purpose and goals of NEPA follow many principles of sustainability. The processes and participants, or stakeholders, to the NEPA process would be the first used for sustainability environmental impact assessments.

The purpose of the NEPA environmental assessment is to reduce environmental impacts on the environment when possible. However, economic considerations specifically and legally drive the decision-making process because they are the overriding value in U.S. society. Nonetheless, the NEPA environmental impact assessment process offers valuable ways to garner important environmental information.

PARTICIPANTS TO THE U.S. ENVIRONMENTAL ASSESSMENT PROCESS

There are four main sets of participants to the NEPA environmental assessment process. The first is the lead agency, which is the agency responsible for EIS preparation and for making the decision on the proposed action. The second set of participants is the EIS team. This group can differ widely from project to project, depending on the range of issues. Generally, an interdisciplinary group of specialists make scientific observations and decisions around these observations. They are usually scientists, engineers, and institutional planners. Each group member is supposed to be fair and unbiased. They are supposed to examine each area as thoroughly as the level of significance of the environmental impact dictates. Group members can be employees of the lead agency or private consultants working under the lead agency, as well as agency staff.

Some advocates of greater inclusion in the environmental decision-making process and as a principle of sustainability criticize the NEPA process at this juncture because there is no citizen or community or environmentalist involvement in the core EIS team. The lack of inclusion, they claim, can cause them to issue an erroneous finding of no significant environmental impacts. The level of significance of environmental impacts is controversial under NEPA. Short-term impacts caused by most construction techniques are not considered significant. The environmental impacts of the mitigation techniques themselves are not considered. Some industry stakeholders argue that simple compliance with air, water, and land environmental laws should be considered part of the mitigation package. Under a regimen of sustainability policy, significance would be tied to potential for irreparable damage to natural systems on which future life depends. There would be little to mitigate if the damage were irreparable, but controversies ensue until the evidence almost reaches levels of species extinction, as occurs in overfishing.

The third group that is part of the U.S. NEPA process is the project proponent. This is usually a private developer or landowner, or sometimes an agency. This group can also be called an applicant or a sponsor. If it needs money, environmental permits, or governmental approval, it probably has to begin the environmental impact process. They have to provide accurate and complete information about the design, construction, and operation of the proposal. They are supposed to share all drawings, feasibility studies, environmental information, and building designs with the lead agency. The lead agency can request more information or explanations of the information provided. Project proponents often complain about the intrusion into business practices and the amount of time an EIS involves. Many claim it scares away potential investors. Sometimes the process of doing an EIS uncovers previously unknown legal liabilities, such as an illegal hazardous waste site. This also affects investors' perceptions of the risk involved with the project. If there are significant environmental impacts that have to be mitigated, the cost of mitigation could be expensive. These contingent liabilities of time-consuming, unknown environmental liabilities and cost of mitigation affect the project proponent. They also affect the NEPA environmental assessment process.

The role of the public as a participant is varied and growing under U.S. NEPA. Generally interested parties of the public are allowed to provide input to the lead agency in the scoping process to narrow the issues and alternatives in a draft EIS, and to review the draft EIS. Many communities have felt excluded by this process. The lead agency generally selects which parts of the public are allowed to participate. Communities want more of a say in developing project alternatives and in designing mitigation schemes. Under NEPA, public review participants include private citizens, Indian tribes, other agencies who have expertise or jurisdiction, and interested parties who have requested notification.

THE NEPA ENVIRONMENTAL ASSESSMENT PROCESS

The process begins when a project proponent submits an application to a federal agency. At this stage of the process, the application may not indicate all the environmental impacts but show preliminary designs and concepts. If it is a state environmental policy act (SEPA) private or local agency actions may not necessarily be covered. Some states require that only lead agencies submit EISs.

After the proposal is submitted, the lead federal agency determines whether the project is categorically excluded from EIS requirements or if it is exempt from them. Both the federal agencies and state governments have these categories. The policy justification for these categorical exemptions is that these activities do not usually cause significant impacts and therefore would not require an EIS. Many environmentalists and sustainability proponents argue, however, that this is not always the case. For example, community development block grants were largely categorically excluded from federal EIS requirements under NEPA. These funds went to many programs in urban areas that had direct and indirect environmental impacts that were not counted or assessed. An adequate platform for a sustainability policy would need to count environmental activities without categorical exclusions or exemptions.

The next step in the U.S. NEPA process is to make a threshold determination of whether there are significant impacts to the environment because of the project. This is a highly controversial and litigated area of the law. Threats to endangered species, wetlands, historic and cultural areas, and controversy itself can trigger a level of significance that requires an EIS. Issues of environmental injustice and racial disproportionality can be a significant impact. If the lead agency is uncertain about whether there will be significant impacts on the environment, they perform an environmental assessment (EA) to determine the potential for significant environmental impacts.

There is substantial community concern and sustainability criticism for this step of the process. Members of the community are not given notice of this EA or even the project application. They have no opportunity for involvement to say what they think the environmental impacts of the proposal would be to them and their environment. The EA is often limited to long-term direct impacts on the study site alone. Some claim the study area is too small to measure ecosystem impacts in most projects, and that the environmental impact study area is manipulated to decrease environmental liability and significance of environmental impacts. For example, in a site where a federal courthouse was proposed, a leaking underground storage tank from an abandoned gasoline station was found. If it were included in the study area, it would have shown a plume of petrochemical pollution from the leaking underground storage tank through the soil, to the water table, and to a nearby river. The cost of cleaning and mitigation of this hazard would be expensive and time consuming. The study area was redefined to exclude the abandoned leaking underground storage sites. Although the community resisted because it wanted to get the site cleaned up, the federal government, and its EIS process, expressly preempts state and local environmental laws.

If the lead agency finds no significant environmental impacts it issues a finding of no significant impacts, or FONSI. This is often the first notice the community receives about a project in its midst. Some communities wholeheartedly endorse any economic development despite environmental consequences. Other communities express shock and outrage at their lack of inclusion on the threshold issue of significant impacts. Environmentally burdened communities are especially sensitive to late notice and exclusion from decisions that directly affect them. Many environmentalists express concern that a more thorough study was not performed and that ecosystem and cumulative effects were not included. Sustainability proponents find this stage of the process lacking because of lack of inclusion and because of lack of ecosystem- or biome-based study. The U.S. EPA's Council on Environmental Quality defined ecosystem as:

> an interconnected community of living things, including humans, and the physical environment with which they interact. Ecosystem management is an approach to restoring and sustaining health ecosystems and their functions and values. It is based on a collaboratively developed version of desired future ecosystem conditions that integrates ecological, economic, and social factors affecting a management unit defined by ecological, not political, boundaries.

> The Twenty-fourth Annual Report of the Council on Environmental Quality.1993. (available on line at ceq.hss.doe.gov/nepa/reports)

If the lead agency does find significant environmental impacts, then under the U.S. NEPA process it issues a notice of intent (NOI) to prepare an EIS statement. The NOI is published in the *Federal Register*. This daily publication is issued to federal depository library institutions. It constitutes public notice of agency actions such as rules, regulations, and EISs.

From here, the public scoping process begins. Scoping is an important part of the U.S. NEPA process because it can determine the actions, alternatives, environmental effects, and sometimes mitigation measures included in the EIS. Different federal agencies involve the public to different degrees in the scoping process. A common complaint, however, is that only those members of the public who agree with the proposal are allowed to participate in a meaningful manner. Many environmentalists and most sustainability proponents would consider the actual scope of the EIS to be too narrow to be applied to environmentally based sustainability approaches. Agencies have the discretion to choose members of the public. Until communities complained about environmentally unjust disproportionate environmental impacts, very few members of the public were engaged in the scoping process. Scoping is done informally and formally by the lead agency, often in close consultation with the project proponent or its consultants. Communities and tribes have been excluded from the process. The lead agency determines the size and scale of the environmentally affected area. This is a controversial decision because it underestimates environmental impacts according to many environmentalists.

PUBLIC INVOLVEMENT IN THE ENVIRONMENTAL ASSESSMENT

Another important area of public involvement under the U.S. NEPA process is public review of the draft EIS. The draft EIS contains all the alternatives to significant environmental impacts. The range of alternatives, including the no action alternative, is often very small. The more alternatives considered the more expensive and time consuming the EIS process can be. The draft EIS can contain important environmental information that could be useful as baseline information in later sustainability assessments.

The public review of the draft EIS is limited under U.S. NEPA processes. Only comments that address certain questions posed by the lead agency are reviewed. Agencies allow little public review from groups that simply do not want the project at all because of environmental impacts. The public is comprised of developers, special interest groups, environmental and economic advocacy groups, individual citizens, and other reviewing agencies. The lack of inclusion at this stage of the NEPA process changed in the late 1990s, primarily because of the political pressure of environmental justice groups and the legal advocacy of environmental groups seeking information from ongoing EIS processes.

EIS: A DONE DEAL OR HONEST ENVIRONMENTAL APPRAISAL?

Environmental lawyers must generally wait until an administrative agency makes a final decision. This is a legal doctrine called "the exhaustion of administrative remedies." The purpose of the doctrine is to leave specialized and complex areas within the expertise of the administrative agency until the complainant has pursued all administrative avenues, giving the administrative agency an opportunity to self-correct. This can be a very long and expensive process, effectively excluding most poor and working people. Interagency appeals processes can take years. The final decision in many cases is not always clear. In the NEPA EIS process, the final lead agency decision is called the Record of Decision and is published in the *Federal Register*. (See www.gpoaccess. gov/fr/.) This may occur long after the EIS is complete. Many environmentalists and communities feel excluded from meaningful participation, leaving their interests unaddressed. Their interests are often in line with sustainability values and approaches, such as preservation of the environment for future generations and the application of the precautionary principle.

Economic Value Prioritized in U.S. NEPA

It is clear that the U.S. NEPA EIS process is laden with a strong economic value directed toward growth. It is very unusual for a project to be denied. Many claim that the environmental mitigation measures claimed in the EIS are unenforceable. The whole decision is not mandatory, merely advisory. From a sustainability perspective, the lack of meaningful participation around ecosystem issues poses a challenge. Although the U.S. EIS process is supposed to examine cumulative impacts, and cumulative impacts are supposed to have a significant impact on the environment, in reality, it is ignored because of the time and resources necessary to evaluate and then mitigate it. The increased environmental scrutiny under assessment procedures, however, may dissuade projects with overwhelming harmful effects from even submitting an application. This is not necessarily the case because of the progrowth assumption of the U.S. EIS process. Nuclear reactors and nuclear waste sites, for example, are subjected to stringent EIS procedures in most cases but eventually are given permission to operate.

Another check on the U.S. NEPA environmental assessment process is the amount of time involved. An EIS can take a long and uncertain amount of time. From the project proponent's perspective, this can decrease investment and ultimately profitability. There is always a certain amount of pressure to streamline the process. This often occurs at the expense of public participation and in-depth environmental analyses. In contested environmental issues, such as timber sales in national forests, industry feels that some public participation is done to help environmental lawyers prepare for their lawsuits once the administrative decision is final.

From an environmental perspective on sustainability approaches, the inability to engage the assessment process judicially, before it is "complete," does not allow for complete and accurate inclusion of all the environmental impacts of a given project. If ecosystem impacts to other systems on which future life depends are irreparably damaged, then the environmental assessment is inadequate for sustainability purposes. It would be awkward to apply the sustainability-based precautionary principle, where threats to natural systems are evaluated before project approvals, to the current environmental assessment process.

U.S. ENVIRONMENTAL PROTECTION AGENCY: CHECKS AND BALANCES?

All federal agencies have environmental impact assessments controlled under NEPA. Most federal agency EISs under NEPA must clear the U.S. EPA and this ensures that the EIS is up to minimal standards. This process occurs in two stages. In the first stage, the EPA evaluates the adequacy of the draft EIS, placing it into one of three categories. The first category, "adequate," describes the impacts of the alternatives developed, and states that no further analysis is necessary. The next category, "insufficient information," means that the draft EIS did not contain enough information for the EPA to assess environmental impacts that need to be avoided. It can also mean that new alternatives were identified that would reduce environmental impacts more than those considered. These new alternatives can come from many sources at this juncture in the process. Communities with late notice of the environmental impact assessment process but strong political power can be one source. Other federal agencies can be another source. It is the decision of the lead agency to decide the scope of alternatives, but the EPA can label the EIS insufficient if it finds an alternative that should be there. This then requires that the final EIS consider the alternative. It does not require that it accept it. The last category of EPA evaluation of the draft EIS is "inadequate." This means the draft EIS does not address significant environmental impacts. It can also mean that new alternatives were introduced that would reduce environmental impacts but were outside the scope of alternatives available in the draft EIS. It requires that the draft EIS be revised and resubmitted for public review as a supplemental draft EIS.

The EPA can also evaluate the environmental impacts of the action in the draft EIS. There are four categories of evaluation. The first category is "lack of objections." The second category is "environmental concerns." This means that the EPA identified environmental impacts that should be avoided. Environmentalists have criticized the EPA for not using this category as a basis for broader environmental concerns, such as sustainability. With environmental concerns, the EPA indicates that it volunteers to work with the lead agency to mitigate the identified environmental concerns. The lead agency does not have to do so. The third category is "environmental objections." Here the EPA identifies significant environmental impacts that must be avoided. In this case, the EPA intends to work with the lead agency to avoid these identified environmental impacts. The last category is "environmentally unsatisfactory." This means that the EPA identified very serious environmental impacts that could endanger the public health and environmental quality. These environmental impacts must be avoided and the final EIS must show it. If the final EIS does adequately deal with these environmental impacts, the matter is referred to the Council on Environmental Quality.

REAL PROPERTY ENVIRONMENTAL ASSESSMENTS

The purpose of this type of environmental assessment is to avoid liability for past and present cleanup responsibilities on the site. The cost of cleanup of land is a large and dynamic factor in a real estate transaction in many industrialized urban areas. A complex and controversial environmental policy question is whether an institution that finances the purchase of a site requiring cleanup is also liable for cleanup costs. Potential buyers and lenders interview contiguous neighbors, sample soil and water, and search public land and environmental records. In the United States, some banks do extensive environmental agency research at the federal, state, and local level. If the land is found to be contaminated, a whole other level of more probing, on-site assessments take place. These are not human health, ecological, or cumulative risk assessments. They are assessments of cleanup liability and generally assume low levels of cleanup limited only to that site. They may assess whether they can divide the property to avoid the contaminated portion of the site. Generally, at this level an environmental assessment must include the costs of remediation in the assessment. A traditional market value appraisal of real property does not offer much useful environmental information about the land. An environmental assessment does provide such information by including actual environmental condition of sites and the cost of cleaning them to the lowest possible standard, without assessment of biological or chemical risks.

ENVIRONMENTAL ASSESSMENTS: ARE THEY ENOUGH FOR SUSTAINABILITY?

The state of environmental assessment is dynamic and growing around the world. The Council of European States requires a post-EIS phase to follow through on the state of any environmental impacts and to see if the promised mitigation is actually working to mitigate the environmental impacts. Its EIS process includes a research and development component so that it may learn lessons from the site and to see how that site fits into its ecosystem. Soon technology will allow us to monitor every site on Earth, including remote locations such as the poles. It also allows for a broad regulatory potential for the environment, something needed for new sustainable policies. Environmental enforcement models rely heavily on assisting compliance. In the United States, if an environmental wrongdoer simply admits to the charge, the fine is reduced

50 to 70 percent and most fines are not collected. Sustainability will require strong enforcement models that will be greatly aided by complete, real-time monitoring of the environment. These enforcement mechanisms will deter environmental behavior that degrades the environment. Technological improvements in environmental monitoring via satellites, cameras, and monitoring stations (staffed and unstaffed) may set the stage for sustainability policies. These improvements vastly increase the number of observations of natural systems, and that knowledge contributes to a burgeoning understanding of the interrelatedness of global ecosystems. Environmental assessments are asked to do more and more, such as include cumulative impacts analysis and ecosystem risk analyses. Emerging from the growing knowledge base available for environmental assessments, the growing application of environmental assessments to more projects, and corporate reporting of the triple bottom line (social, economic, environmental, *See also* **Volume 2 Business and Economics** is a new emerging "sustainability" assessment.

The amount of growth of environmental knowledge is rapidly increasing, spurred by the knowledge that natural systems have limitations and that humans have a large effect on these natural systems. This rapidly increasing knowledge base of environmental limitations relies heavily on environmental assessment. That assessment, however, depends on the purpose for which it is used.

References

Clark, William C. et al. 2006. *Linking Knowledge with Action for Sustainable Development: The Role of Program Management.* Washington, DC: National Academies Press.

Holder, Jane. 2009. *Taking Stock of Environmental Assessment: Law, Policy, and Custom.* London, UK: RICS Books.

Sustainability Assessments and Audits

Although the environmental assessment process offers the best foundation for a sustainability assessment, its application to policies and practices increases under sustainability. Determining roles for sustainability assessment in the policy process is one of the first policy developments.

GENERIC SUSTAINABILITY AUDIT FOR MOST INSTITUTIONS

Most institutions and organizations have policies and practices that contribute to environmental impacts, sometimes in an unsustainable manner. While environmental audits were developed to help industry comply with environmental laws, they did not reach broad application to non-industrial sectors such as residential or institutional uses like schools. For an audit to be useful as a sustainability audit, and for it to be a building block of a sustainability policy, it must apply to all possible human impacts because all impacts can contribute to increasing cumulative impacts. Sustainability audits are in a dynamic state of growth. Currently little financing or law requires them, but they are the tool that will implement the sustainability missions in international documents and agreements.

Most institutional sustainability audits cover all environmental impacts from the organization. Organizations can differ greatly in size, function, mission, and environmental practices. Churches, municipal offices, schools, state offices, and neighborhood organizations are a few examples of nonindustrial institutions and organizations. Many of these organizations are interested in being sustainable, or at least learning more about it. Some may not be able to afford a "sustainability audit" or fear that it may reveal environmental liabilities. Some environmental advocacy groups are promoting the idea, and some consulting groups have begun to expand from the usual corporate practice into municipal and state clients.

PRIMARY FOCUS OF GENERIC SUSTAINABILITY ASSESSMENT

The primary focus of a generic sustainability assessment of an organization is energy usage, waste production, water usage, paper usage, staff and human resource policies, travel practices, and purchasing practices. The ultimate goal is to know the organization's ecological footprint. Ecological footprint analysis measures the amount of renewable and nonrenewable ecologically productive land area required to support the resource demands and absorb the wastes of the organization.

In terms of energy usage, the goal of a sustainability assessment is to reduce consumption, and certainly any wasteful or inefficient practices. This can vary greatly depending on the climate. Energy conservation is often a sustainable practice that saves money for most consumers. Energy should also be procured from renewable sources, even if they are more expensive in terms of cash cost. This may require some research and investigation because some renewable sources can be on site. Solar panels, for instance, could be added to structures housing the organization. Sustainability assessments also emphasize the use of energy-efficient equipment and energy infrastructure. Sensor-activated appliances such as lights, escalators, and people movers are prioritized. The use of passive systems such as awnings and use of cross-ventilation from open windows is also assessed. Designs and practices that use natural light as much as possible are also part of an audit.

In terms of waste overall, the sustainability audit focus is to reduce use, reuse, and to recycle all wastes. Waste issues are important in a sustainability audit. In some industrial ecology approaches to sustainability, the idea is to close the loops of waste production. This often requires a search for markets for what was once disposed of as waste. Life cycle analysis of buildings can show areas of waste and heavy environmental impact as well as potential markets for "waste" and places for new products.

For example, one new product is replacing drywall. Drywall, or plasterboard, was invented in 1917 and is produced by baking grounded gypsum rock at 500 degrees. This process emits about

20 billion pounds of greenhouse gases a year and uses a large amount of energy. About 136 million tons of construction and demolition wastes disposed of in the United States each year, although that number is probably higher because of underreporting. Much of it is drywall. There is now a new drywall made of construction and other wastes that chemically bind without heat. The production process uses one-fifth less heat than the 1917 method. It does not use starch so that termites and mold are less of a problem. Lifecycle analysis shows that 48 percent of carbon dioxide emissions in the United States came from buildings in 2006, and probably a large part of that was drywall production and wastes.

In many organizations, paper waste is an issue because the use of computers and printers increased demand for paper. Many offices now try to go as paperless as possible to reduce paper waste. Two-sided copying and printing is encouraged in organizations with a sustainability ethos. Paper recycling is highly encouraged and is made as convenient as possible. The type of paper can be important. Using paper for printing, paper towels, and toilet paper from recycled or sustainable sources is also assessed. In terms of other wastes, as much of it as possible should be recycled, including toner cartridges, aluminum, glass, plastics, and packaging. Obsolete objects of technology, such as older computers, should be recycled or donated if possible. Food implements should be reusable, and organic waste should be composted.

The environmental impact of the institution on water is very important in a sustainability audit, especially if water is a threatened natural resource in that location. Here a sustainability audit would seek to minimize water use and to pretreat it before it enters natural water systems. Some ways to reduce water usage is to use motion-sensitive faucets. Motions sensors can also be used on hand air dryers to save energy and paper. The use of toilets that provide two different types of flushes—one for solid waste and the other liquid waste—is another way to reduce water usage. Sustainability audits also examine the land around the organization in terms of landscaping practices. Generally, the use of indigenous species requires less fresh water than the traditional grass lawn. Environmentally sensitive irrigation and pest management programs can also lessen environmental impacts on natural water systems. The choice of cleaning materials and soaps is important because many of these are simply washed down the drain. The use of nonphosphorous soaps is another way of reducing impacts. Impacts on water resources can also be reduced by using gray water. This is water directly from the roof or other catchment area, and water sinks and drains from cleaning and bathing. Gray water can be used for landscape water maintenance, toilets, and external washing. Knowledge of the complete chemical content and biodegradability of all indoor and outdoor cleaning materials is essential for accurate knowledge about the ecological footprint of the organization. Many organizations in the United States file a material safety data sheet with the Occupational Safety and Health Administration (OSHA).

MATERIAL SAFETY DATA SHEET (MSDS)

This chemical reporting system is part of the hazard communication standards. These standards were developed by OSHA. The purpose of these standards is to protect workers from chemical hazards while in the workplace. Some of these chemicals are used in homes and other nonwork locations. By law, the MSDS plainly lists the chemical composition of the substance used, its trade name, and the name of the manufacturer, the hazards from the substances, and precautions workers should take to avoid the hazards. MSDS are prepared by the manufacturer of the chemical or chemical product. MSDS are required to be kept on site and made available to the public and to workers by the employer. MSDSs provide a valuable source of information for sustainability audits.

A sustainability audit useful for generic application to most institutions and organization also examines the organizational process and people. Here there are usually three areas of emphasis: staff and human resources, transportation, and procurement.

In terms of staff and human resources, the sustainability audit will assess the understanding and awareness of issues that the particular assessment views as relevant to sustainability in that organization. Because sustainability views can range from sustainable quarterly profits for a corporation to radical swings in values and consumptive behaviors by communities, the assumption of the sustainability assessment should be stated. An assessment could examine efficiency in production as part of a sustainability audit. Generally, a generic sustainability audit will look for environmental information points and forums, communication around environmental issues like procurement and recycling, specific positions such as sustainability coordinator, and other social activities. Continuing education programs in professions and trades that encourage staff to learn about sustainability is also part of this aspect of the sustainability audit.

Transportation and staff travel can dramatically increase the carbon use of an organization. Most alternatives to private vehicles are encouraged, such as walking, bicycling, and mass transit. The provision of storage for alternative transportation and showers for cyclists is also assessed. Telecommuting, or working at home via the Internet and other communication systems, is also encouraged in the audit. In organizations with large fleets of vehicles, the sustainability audit will examine car pools, purchasing of carbon dioxide emission offsets, and use of fuel-efficient and low-emission vehicles.

Most institutions and organizations have to purchase goods and procure services. In terms of a sustainability audit, a lifecycle impact assessment for all purchased goods is evaluated. This means that how the product is manufactured, used, and disposed of is included in the computation of the ecological footprint. Recycled paper and other products are highly prioritized. In examining the supply chain, the purchaser should support local suppliers if transportation costs of distant suppliers increase the environmental impact. Purchasing supplies in bulk can also lower transportation impacts on the environment by decreasing the number of delivery trips. In terms of office supplies, the use of low volatile organic chemical emissions from paint, carpets, glues, and sealants is preferred under a sustainability assessment.

A generic sustainability assessment for noncorporate institutions and organizations is a large, dynamic, and necessary step for sustainable policy development. Corporate institutions, college campuses, and communities should all do their sustainability assessment. They all have different purposes and functions because they are used by different stakeholders.

References

Devuyst, Dimitri et al. 2001. *How Green Is the City?: Sustainability Assessment and the Management of Urban Environments.* New York: Columbia University Press.

Gibson, Robert B. et al. 2005. *Sustainability Assessment: Criteria and Process.* London, UK: Earthscan.

Hitchcock, Darcy, and Marsha Willard. 2006. *The Business Guide to Sustainability: Practical Strategies and Tools for Organizations.* London, UK: Earthscan.

Schaltegger, Stefan et al. 2006. *Sustainability Accounting and Reporting.* New York: Springer.

CORPORATE SELF-REPORTING OF SUSTAINABILITY

Corporate self-reporting of sustainability is emerging from a corporate practice of the triple bottom line of environment, social, and economic measures. Corporations have experience gathering information from sustainability indicators they use and develop, then developing information systems that can be used for internal audits. The use of this information for external third-party review or auditing is necessary for sustainability. To audit information means that it is verifiable.

SUSTAINABLE ASSESSMENTS OF TRADE POLICIES

The strong interest in sustainability in the international community combined with a global concern about all ecosystems pushes sustainability assessments to trade policies. One of the environmental concerns is that as global markets open and expand, some natural resources will be irreparably damaged. Some international environmental organizations advocate for the use of sustainability assessments to shape trade and investment agreements so that these support sustainable development. Some of these types of trade sustainability assessments go beyond environmental measures and include social and equity measures.

Different standards are used by various industries. One that they prefer is the Global Reporting Initiative (GRI), a large multistakeholder network of experts around the world. They use the GRI guidelines to report, access information in GRI-based reports, or contribute to develop the reporting framework.

INTERNATIONAL ORGANIZATION FOR STANDARDIZATION: ISO STANDARDS FOR SUSTAINABILITY

Another set of standards used internationally by industry is being developed by the International Standards Organization. The International Organization for Standardization is the world's largest developer and publisher of International Standards. It has 157 member countries and is itself a nongovernmental organization. It developed a series used in environmental audits called the ISO 14001, although it was not developed specifically for sustainability. Most ISO products are developed for a specific product or process. ISO 14001 designs a process-based approach to environmental management of an industrial plant. It does have the ability to monitor a wide range of environmental issues, and offers a potential platform for a sustainability assessment. The ISO 14001 process can be applied generically. It does not list any performance goals or criteria and relies on internal audits.

Some industry and corporate sustainability audits are done by consultants who offer sustainability auditing as a service for hire. Currently, most examine the carbon footprint of the organization, the operations and culture to see if sustainability ideas and policies are proactively managed, the procurement of supplies for operations and for product, and socioeconomic impacts. This type of internal, industry-focused sustainability audit identifies employment and income generated from building activities as socioeconomic impact.

Industries and the corporations and partnerships that manage them have been at the cutting edge of environmental regulation; however, they are not the only source of environmental impacts. The sheer volume of human population increases will generate large impacts on natural systems. The scope of environmentally regulated activities will expand under sustainability to areas of society other than big industry. It will spread to reach smaller industries, and then municipal emissions, military emissions and wastes, and other waste streams that impact the environment. Knowledge of the environmental impacts of development provides much of the impetus toward sustainability.

U.S. COLLEGE CAMPUSES AND SUSTAINABILITY ASSESSMENTS

As centers for advanced learning and education, colleges and universities are places for leadership in environmental policy. This is also true in the area of sustainability. They are also institutions that can have a large impact on the environment via their physical plant. Many colleges and universities have large research programs that use dangerous chemicals. Many are exempt from increasingly stringent right-to-know laws at the federal, state, and local levels. Many campuses have large populations of childbearing adults, for whom exposure to environmental stressors is a concern. As knowledge about the environment develops, one of the places it develops first, and where it is combined with other recent advances in other fields of knowledge, is the college or university. This environmental knowledge gives extra impetus to campus environments. When combined with the enthusiasm and intellectual energy of college students, staff, and faculty, this burgeoning area of environmental knowledge seeks expression in the assessment of the college or university itself. As noted by several legislators who sponsored the Higher Education Sustainability Act:

> Colleges are a natural breeding ground for the kind of innovation we need to move to new, environmentally friendly energy sources. Our young people know the stakes. They know that developing sustainable energy programs will affect their lives, their economic well-being, and the planet they are inheriting. These grants will help college students take the reins of the movement to make energy last longer and have less an impact on the environment.
>
> Senator Patty Murray, original Senate Sponsor
> of the Higher Education Sustainability Act.

> What better way to promote sustainability than to encourage our institutions of higher learning to create academic programs to teach its concepts, and to implement sustainable practices themselves. Society will reap the benefits of the excellent return on investment gained by educating students in sustainable practices.
>
> Congressman Vernon Ehlers, original House
> of Representatives co-sponsor.

The National Wildlife Federation and Princeton Survey Research Associates International surveyed U.S. higher education programs on the topic and practice of sustainability in 2008. They examined 1,068 institutions of the approximately 3,000 similar institutions in the United States. Academic letter grades (A through D) were awarded for collective, national performance on environmental literacy, energy, water, transportation, landscaping, waste reduction, environmental and sustainable goal setting, and more. Their report analyzes collective trends in three areas—management, operations, and academics. A total of 27 percent of U.S. colleges responded to the 2008 survey. It is the nation's largest study to date created to gauge trends and new developments in campus sustainability.

Colleges were ranked according to their responses to 18 question sets that assessed the institution's commitment to sustainability. In this way, it was a rough sustainability assessment, or at least the beginning of one.

The 2008 report card on academic institutions reveals that many are starting to put more resources into energy conservation practices, and more are setting goals to reduce carbon emissions. In all, 65 percent of respondents had a written commitment or plans to develop one to sustainability. Many more colleges have green landscapes, energy efficiency programs, and recycling programs. More than 90 percent of respondents had hired or planned to hire an energy conservation manager. About half of the respondents had a position for a green purchasing coordinator. Willamette University, for example, was part of the list of the top 40 academic respondents that are "exemplary schools for environmental or sustainability goal setting" and "exemplary for supporting and evaluating faculty on environmental or sustainability studies."

Other organizations, such as the Association for the Advancement of Sustainability in Higher Education (AASHE) are developing their sustainability assessments of college environments. AASHE is developing a sustainability assessment system called STARS (sustainability tracking, assessment, and rating system). It covers three dimensions of sustainability—social, economic, and environmental.

College campuses are places of energy, enthusiasm, and focus, especially around issues of sustainability. They are also places of privilege. As

Willamette University Campus Sustainability Plan

Willamette University is one of the leaders in the United States in the area of sustainability, a philosophy embodied in the institution's motto "Not unto ourselves alone are we born." It was ranked first in the nation out of 240 colleges for engaging in the most sustainable activities. The university has a Center for Sustainable Communities, a strong commitment to energy conservation and efficiency, pursues transportation alternatives, pursues LEEDS certification in its new campus buildings, develops environmentally friendly landscaping practices, and provides a sustainability orientation to new faculty and staff. It includes several sustainability courses in its curriculum, including sustainability and the law, a course that was first taught in U.S. law schools by one of this book's authors, Professor of Law Robin Morris Collin. Professor Robert Collin, the other author of this book, is also employed at Willamette University as senior research scholar in sustainability. The approach of Willamette University was motivated by the faculty and students, and encouraged and facilitated by President Lee Pelton.

The process at Willamette was inclusive and time consuming, but not especially expensive. People volunteered their time and effort. Occasionally a small grant from the administration paid for food or rental space. A sustainability council was formed and met frequently. Several intensive sustainability retreats were held, and these continue, providing an environment where students, staff, and faculty discuss issues of sustainable economics, equity, and environmentalism. There were few large outside grants or expensive consultants. The governing board is fully apprised of all developments and is supportive of these efforts. Virtually everyone on campus is aware of sustainability and has an opportunity to participate in sustainability activities.

Other universities are developing innovative sustainability policy with their own types of assessments. Examples include New Jersey's Campus Sustainability Snapshot (www.njheps.org/assessment/guide.htm), Berkeley Sustainability Assessment (sustainability.berkeley.edu/assessment), and University of California at Santa Barbara Campus Sustainability Plan (sustainability.ucsb.edu/plan).

a result, many campuses have their own sustainability assessment but do not include many equity or transparency measurements or observations of their ecosystem outside their own institution. There are also issues of inclusion of all stakeholders, such as maintenance and athletic staff. Sustainability measures included transparent environmental measures that can be observed by others. In many of the pioneering sustainability assessments, the selection of indicators or benchmarks is a major part of the process. U.S. universities are often not subject to right-to-know laws and generally do not voluntarily report chemical emissions or pollution. Getting these institutions to a level of environmental disclosure suitable for sustainable development is a major challenge. This is the case for many of the first-generation college campus university assessments. Other organizations are also developing their own campus sustainability assessments.

The youthful energy of the campus clients, the students, drives a dynamically changing and growing approach to sustainability. This growth to date is uneven. The most growth is in the operation of facilities responsible for the day-to-day operations of the campus. Traditional programs of energy and water conservation take hold first in usual ways. Insulating rooms against heat loss, programmable thermostats, and monitoring water usage are basic beginnings. Some college campuses are taking stronger steps toward sustainability such as pursuing alternative energy sources, reclaiming gray water, and reducing paper waste by going to a paperless work environment. Some colleges are planning to reduce their carbon dioxide emissions and to increase voluntary efforts such as recycling. Other steps toward sustainability falter. Many students, staff, and faculty commute to work with cars. Mass transit, alternative energy vehicles, and car pooling do not yet work with the college population, although many colleges are currently exploring these options.

One area that lags behind others in college sustainability programs is the political economy of knowledge around sustainability. Colleges are divided up into disciplines that traditionally determine who they hire and retain. When academic employees break academic ranks and goes outside the traditional disciplinary boundaries of their science, humanity, or social science, they are generally not rewarded. Some disciplines require publication in their respective discipline. Many of these publications are peer reviewed. They strongly maintain the disciplinary boundaries through which articles are accepted or rejected. In this way they determine which academic employees stay or are rewarded. Sustainability is not a separate discipline, but considered a meta-discipline at best. Sustainability courses, majors, and programs show continued decline as colleges become more competitive along traditional disciplinary lines. About half of all colleges have a major or minor in environmental or sustainability areas, down from about two-thirds in 2000. Even those colleges with an environmental major or minor make it an environmental sciences focus that may not be broad enough for sustainability concerns.

Tension between college physical plants, the needs and desires of students, and the political boundaries of disciplines is increasing because the global community is concerned about sustainability. The world needs leaders of the future to be knowledgeable about the challenges they face. One of the main challenges is how to manage sustainable development in the face of climate changes and potential global scarcity. These issues engender global concerns about equity, inclusions, and responsible environmental action based on knowledge.

Universities in other nations are also pursuing the development of a sustainability assessment. For example, Canada developed the Campus Sustainability Assessment Framework. The "Greening the Ivory Towers Project" uses Canada's Campus Sustainability Assessment Framework.

Reference

Association for the Advancement of Sustainability in Higher Education, www.aashe.org/.

NEW LAW? HIGHER EDUCATION SUSTAINABILITY ACT

On August 1, 2008, Congress passed the Higher Education Sustainability Act as part of the Higher Education Opportunity Act (HR 4137). In terms of campus sustainability, HR 4137 contains two important provisions. First, it provides authority to convene a national summit of higher education sustainability experts, federal agency staff, and business leaders to identify best practices in sustainability, as well as opportunities for collaboration to expand sustainable operations and programs. It also sets up a University Sustainability Grant program that authorizes the secretary of education in consultation with the U.S. EPA, to offer competitive grants to colleges and universities to establish sustainability research programs.

The House and Senate (which passed their new Higher Education Act, S. 1642, in August 2009) are currently conferring on the final bill. This conference committee meeting will decide which of the 40+ new programs, including the HESA (Higher Education Sustainability Act) program, make it into the final higher education bill. HESA amends the Higher Education Act to authorize a new $50 million grant program at the Department of Education that will annually support between 25 and 200 projects at higher education institutions and consortia/associations. It is very likely that many of these projects will develop and refine sustainability assessment tools and indicators.

MITIGATION OF ENVIRONMENTAL IMPACTS UNDER SUSTAINABILITY ASSESSMENTS

Like most other environmental assessments, most sustainability assessments will contain mitigation measures to offset impacts on the environment. Mitigation under a sustainability scenario may have to prevent irreparable damage to natural systems. Under most current environmental assessments, mitigation is used only to ameliorate, or diminish, the environmental impacts. This standard may not be strong enough to protect damage to natural systems without application of the precautionary principle, a principle of sustainability where risk of irreparable damage to natural systems is first assessed. Also, environmental mitigation under typical U.S. environmental policy suffers from a lack of enforcement. Compliance is sought, but environmentally wrongful behavior is seldom deterred through punishment under current regulatory protocols. Some environmental mitigation may have significant environmental impacts itself, but under NEPA, mitigation itself does not have to be mitigated. For example, if the straw used to mitigate or prevent erosion, a significant environmental impact, from a construction site is itself an environmental hazard, it remains unaccounted in the process but not in the environment.

PROBLEMS IN MITIGATION UNDER SUSTAINABILITY

Any level of environmental mitigation is costly in terms of time and money. The levels of environmental mitigation alone that are necessary under sustainability would be resource intensive and needed over time. The business-as-usual compliance model used under traditional models of environmental assessment would be inadequate for a sustainability assessment of the same project. Ideally, impacts on natural systems need to be known, the health of the ecosystem of the biome needs to be known, and the impacts of the project over the life of the project should be known. This is an expensive level of mitigation for any one project. Some institutions, such as colleges and corporations, have the resources in terms of time, money, and people power to engage in sustainability assessments. The resources necessary for such an intensive assessment, including the conditions of more stringent environmental oversight of all mitigation, are a barrier to changing nonsustainable patterns of behavior.

According to the Intergovernmental Panel on Climate Change (winner of the Nobel Prize along with Al Gore):

> Making development more sustainable by changing development paths can make a major contribution to climate change mitigation, but implementation may require resources to overcome multiple barriers. There is a growing understanding of the possibilities to choose and implement mitigation options in several sectors to realize synergies and avoid conflicts with other dimensions of sustainable development.

IPCC, 2007 Internationally there is high agreement and much evidence for the important role of mitigation as part of climate change and its part in sustainable development. Mitigation of climate impacts is part of sustainable development. Mitigation technologies and practices differ by industrial sector and are reliant on the current state of technology and environmental knowledge.

Sector	Key mitigation technologies and practices currently commercially available.	Key mitigation technologies and practices projected to be commercialized before 2030.
Energy Supply [4.3, 4.4]	Improved supply and distribution efficiency; fuel switching from coal to gas; nuclear power; renewable heat and power (hydropower, solar, wind, geothermal and bioenergy); combined heat and power; early applications of CCS (e.g. storage of removed CO_2 from natural gas)	Carbon Capture and Storage (CCS) for gas, biomass and coal-fired electricity generating facilities; advanced nuclear power; advanced renewable energy, including tidal and waves energy, concentrating solar, and solar PV.
Transport [5.4]	More fuel efficient vehicles; hybrid vehicles; cleaner diesel vehicles; biofuels; modal shifts from road transport to rail and public transport systems; non-motorized transport (cycling, walking); land-use and transport planning	Second generation biofuels; higher efficiency aircraft; advanced electric and hybrid vehicles with more powerful and reliable batteries
Buildings [6.5]	Efficient lighting and daylighting; more efficient electrical appliances and heating and cooling devices; improved cook stoves; improved insulation; passive and active solar design for heating and cooling; alternative refrigeration fluids, recovery and recycle of fluorinated gases	Integrated design of commercial buildings including technologies, such as intelligent meters that provide feedback and control; solar PV integrated in buildings
Industry [6.5]	More efficient end-use electrical equipment; heat and power recovery; material recycling and substitution; control of non-CO_2 gas emissions; and a wide array of process-specific technologies	Advanced energy efficiency; CCS for cement, ammonia, and iron manufacture; inert electrodes for aluminium manufacture
Agriculture [8.4]	Improved crop and grazing land management to increase soil carbon storage; restoration of cultivated peaty soils and degraded lands; improved rice cultivation techniques and livestock and manure management to reduce CH_4 emissions; improved nitrogen fertilizer application techniques to reduce N_2O emissions; dedicated energy crops to replace fossil fuel use; improved energy efficiency	Improvements of crops yields
Forestry/forests [9.4]	Afforestation; reforestation; forest management; reduced deforestation; harvested wood product management; use of forestry products for bioenergy to replace fossil fuel use	Tree species improvement to increase biomass productivity and carbon sequestration. Improved remote sensing technologies for analysis of vegetation/soil carbon sequestration potential and mapping land use change
Waste [10.4]	Landfill methane recovery; waste incineration with energy recovery; composting of organic waste; controlled waste water treatment; recycling and waste minimization	Biocovers and biofilters to optimize CH_4 oxidation

FIGURE 1.17 • Key Mitigation Technologies and practices by Sector. Sectors and technologies are listed in no particular order. Nontechnological practices, such as lifestyle changes, which are cross-cutting, are not included in this table. IPCC Climate Change 2007: Working Group III Report "Mitigation of Climate Change, Summary for Policymakers," p. 14.

Mitigation recommendations are fraught with high levels of uncertainty. Risks of climate feedback, such as between carbon cycles and additional climate changes, are not included in most projections. There is a strong chance that the emission levels necessary to reach a lower level are underestimated. The IPCC recommends short-term mitigation strategies that can help policymakers move toward long-term emission goals. This may help control the costs and efficiency of a particular mitigation approach. Sometimes abrupt changes to an ecosystem via a mitigation approach can be environmentally degrading. The mitigation approach will need continual assessment.

A large area of uncertainty is the extent industrialized nations are locked into their current, high emission processes, and the extent developing nations will use these development approaches. Much depends on the financial, cultural, and physical infrastructure of the individual nation. The choice of a mitigation policy is dependent on the infrastructure of a country around these factors. It is also dependent on what global temperature and level of greenhouse gas emission is for the world policy goal for sustainability. This is partially dependent on whether large-scale impacts such as deglaciation of the Greenland ice sheet, degradation of ecosystems and coral reefs, and constraining species extinction are a world goal. Other world sustainability goals such as climate changes that increase flooding, drought, loss of polar ice sheets, effects on indigenous peoples engaged in subsistence lifestyles, extreme weather events, and net decreases per capita of food production are also factors in choice and development of mitigation approaches. The actual sensitivity of a climate to change and the delay in effects on the climate once a mitigation policy is in effect are also factors.

Some of these factors are local; others are regional or global. Many of these factors interact with each other in currently unknown ways.

There is international agreement that mitigation policies should be developed without regard to scale. As the IPPC observes about "development paths":

> National circumstances and the strengths of the institutions determine how development policies impact Greenhouse gas emissions. Changes in development paths emerge from the interactions of the public and private decision processes involving government, business and civil society, many of which are not traditionally considered as climate policy.

When the IPCC is discussing the synergy between climate change and sustainable development, it focuses on specific industrial sectors. From its research it concludes that macroeconomic policy, agricultural policy, multilateral bank lending, insurance practices, electricity market reform, energy security, and forest conservation could effectively work together, or synergize their efforts towards sustainability. The IPCC also focuses on the adaptability of natural systems to climate changes. Tying mitigation policies to climate adaptation policies is also seen as synergistic. As the IPCC notes:

> Making development more sustainable can enhance both mitigative and adaptive capacity, and reduce emissions and vulnerability to climate change. Synergies between mitigation and adaptation can exist, for example in properly designed biomass production formation of protected areas, land management, energy use in buildings and forestry.

The IPCC fourth report details these findings with much more specificity. It provides clear policy parameters for sustainable development, and shows how systems of nature on which future generations will depend will be impacted by current knowledge and choices. The choice of mitigation policies will be key. In places that adopt the precautionary principle, mitigation may be required as a condition of development. Mitigation can mean many different activities in an international context. It can mean just saying no to development. Much depends on whether mitigation is to be full and complete mitigation, restorative mitigation, or just partial mitigation. The policy timeframe for mitigation is also important. Is it temporary, long term, or permanent mitigation? In many ways the term *mitigation* is really the substantive policy choices a community or nation must make to affect global warming and climate change.

Reference

IPCC. 2007. "Summary for Policy Makers." In *Climate Change 2007: Mitigation*. B. Metz, O. R. Davidson, P. R. Bosch, R. Dave, and L. A. Meyer, eds., 1–25. New York: Cambridge University Press.

Community Sustainability Assessment

Communities are interested in sustainability. They may be in the best position to monitor small to medium ecosystem impacts. Larger, regional ecosystems can span political jurisdictions that often define communities. Several community-based sustainability assessments are starting to emerge. They share some common characteristics in application and often involve extensive surveys and questionnaires.

WHEN ARE THE ECOLOGICAL ASPECTS OF COMMUNITY LIFE SUSTAINABLE?

Is a sustainable utopia possible? In many places, communities are deeply connected to the place in which they live, and this is a good foundation for developing sustainable approaches. Environmentally, communities know the seasonal variations in climate, the history of land uses, and the sites of the greatest environmental degradation. In a sustainable community, the ecosystem's dynamic boundaries, vulnerabilities, and natural rhythms are known and monitored. In a sustainable community, humans do not degrade natural systems irreversibly, and sustainable communities apply the precautionary principle to all land use decisions. Natural systems and processes are understood and incorporated into public policy and private behavior in a sustainable community. In an ideal sustainable community, the food comes from local sources and is organic and healthy. The built environment should be designed with an emphasis on minimizing any environmental impacts. This includes materials for buildings, roads, and paved areas. In ideal sustainable community transportation, impacts would be reduced through road design policies, work flexibility, mass transit, pedestrian and bicycle accommodation, and telecommuting. Waste generation is kept to a minimum through policies of recycling, reusing, and reducing consumption. In a sustainable community, the water must be protected as a highly valuable ecosystem resource. The bioregional hydrology should be known and monitored. Energy usage should come from locally renewable sources such as the sun, the geothermal heat if available, wind, or sometimes small hydropower sources. An ideal sustainable community would be healthy, economically vibrant, and an integral part of the ecosystem.

Reference

Davenport, John and Julia L. 2006. *The Ecology of Transportation: Managing Mobility for the Environment*. New York: Springer.

WHAT DO COMMUNITY SUSTAINABILITY ASSESSMENTS ASSESS?

The aspects of the ecology that community sustainability assessments examine are a sense of place in the community and the location and scale of the built and unbuilt environment. This directs many assessments to the parts of the ecosystem that can be preserved with immediate direction action, as well as parts of the ecosystem that are in danger of irreversible degradation. Areas with native habitat are especially valuable, and areas of sensitive ecosystems especially noteworthy in a community-based sustainability assessment. The actual physical environment inclusive of roads, bridges, fences, buildings, and paved areas is examined for environmentally based ecosystem impacts. How food and waste move through a community is important in terms of measuring ecosystem impacts. Levels of consumption, recycling, and strong local environmental advocacy are other aspects of a community-based sustainability assessment. Use and efficiency of energy sources, water use and efficiency, and air pollution control are other aspects of an assessment. Sustainability assessments also examine the flow of information, the acceptance of diversity, the conflict resolution system, and major institutions in the community. They can also include an examination of spiritual values as they relate to sustainability and the environment.

On a more practical level community sustainability assessments require the local government to prioritize them. They need to know which agencies, commissions, boards, and departments can devote their employees to the tasks. They need to assess whether they have anyone in house that is competent to do any environmental assessment. If they do not, they need to assess whether they can create partnerships with local businesses, universities, churches, or schools to do the necessary work. They need to assess, and perhaps inspire, the community to become a meaningful participant in the process. To do all these tasks, some communities engage in a process of "visioning."

The process of visioning for purposes of a sustainability assessment asks some specific and hard questions. Some of these relate to the values of the community. Many communities have different values, and some people may not have environmental values but will have public health values. Visioning asks what they value most. What parts of where they live, work, play, learn, and worship would they want to preserve for future generations. What current programs or policies do they like and how could they be applied to a community assessment of sustainability. What are their ideas about improvement and how would they be involved in these ideas if implemented.

From these basic visioning processes, the community assessment of sustainability could focus on a variety of sustainability issues depending on their priorities. A focus on renewable energy sources would examine solar access, ways to minimize mechanical heating and cooling, the feasibility of windmills, and energy conservation. A focus on reducing environmental impacts on water could examine home and town water management systems such as grey water, water conservation, and efficient water irrigation and water harvesting systems. A community interested in more efficient waste management could begin a home and/or community composting program. Communities envisioning a more sustainable transportation system could make alternative forms of transportation more accommodating with bicycle transportation planning, shared cars, lower parking rates for electric cars, higher tolls on high air pollution days, pedestrian planning, and alternative forms of mass transit. A community envisioning a healthy construction process could change land use and building codes to require low toxic construction products and processes, certifiably sustainable wood products when wood is used, and high levels of energy efficiency. It could also encourage the construction of local energy sources in residential, commercial, and industrial development such as solar panels or windmills.

Visioning is a proactive process that depends on an assessment of the current sustainability of the community. That is, in part, it is dependent on the will of the community and its priorities for the future.

References

Bell, Simon, and Stephen Morse. 2008. *Sustainability Indicators: Measuring the Immeasurable?* London, UK: Earthscan.

Bunnell, Gene. 2002. *Making Places Special: Stories of Real Places Made Better by Planning.* Chicago: Planners Press, American Planning Association.

Helming, Katharina, Marta Perez-Soba, and Paul Tabbush. 2008. *Sustainability Impact Assessment of Land Use Changes.* New York: Springer.

Kasemir, Bernd, and Jill Jager. 2003. *Public Participation in Sustainability Science: A Handbook.* Cambridge, UK: Cambridge University Press.

Roseland, Mark. 2005. *Toward Sustainable Communities: Resources for Citizens and Their Governments.* Gabriola Island, British Columbia: New Society Publishers.

ENVIRONMENTAL AUDITS

Environmental audits as traditionally applied by industry are used to determine necessary environmental permits, finds energy savings, and reduce wastes. There is a controversy about whether they are privileged, or allowed to be kept confidential. Many U.S. states have laws that protect industry environmental audits. Some argue that if they were not confidential, industry would not do them. Others argue that all environmental impacts are important. From a sustainable development perspective, the traditional environmental audit is inadequate because the environmental information is not transparent, monitored, or placed in an ecosystem context.

Ideally, environmental audits assess the environmental condition and impacts of property, processes, and facilities in a systematic and objective manner. Traditionally, they are used as a management tool to follow environmental laws and regulations and correct or mitigate any problems. They can be used to find cost saving in energy use and waste management. As used with sustainable development, environmental audits can be pushed to evaluate the potential for any activities that would irreparably harm the environment for future generations.

Environmental auditing began when the U.S. EPA began, in the early 1970s. It has evolved into a set of specialized management tools as applied to industry. Environmental audits are generally done internally but as they have become specialized, they have been done by consultants. Currently, there is no legal rule that requires an external environmental audit, although a court decision could require it in a specific case. As the demand for sustainable development increases, more stakeholders from the community and from environmental groups want independent, external audits that go beyond minimal compliance with environmental laws. It is likely that environmental audits will evolve in this direction, such as the LEEDS evaluations for sustainable building construction.

Industries have different reasons for engaging in environmental audits. Mainly, they are done to meet the requirements of environmental laws and rules. Increasingly, they are done to examine their liability for cleanup of polluted sites. As pollution cleanup costs increase, industries must disclose them to potential investors as contingent liabilities, especially if they are to be publicly traded. Environmental audits can be specialized. The basic compliance environmental audit simply ensures the industrial facility is following all the rules and laws. There are specific environmental audits of property transfers to examine the extent of liability for pollution and the requirement of due diligence to discover or reveal pollution in a real estate transfer. There is an environmental audit of internal operating systems that focuses on environmental risk management. Hazardous materials that must be followed from cradle to grave in their treatment, storage, and disposal have their own specific environmental audit. These environmental audits can be combined, and an overall audit is used by industry to determine what, if any, environmental liabilities need to be voluntarily disclosed.

Environmental audits can be mandated by law or court decisions, or they can be voluntary. Traditionally, they have been used in industry to avoid environmental controversies. Industry strongly prefers that they be voluntary; however, more communities are considering their own environmental audits of industry, and of the operations of their own community. Environmental audits of complex organizations, such as some industries, universities, and government agencies, usually require trained professional people skilled in water, air, energy, and waste evaluations. Under sustainable development, there is a need to have independent auditors that pull together objective facts.

U.S. EPA Environmental Audit Rules

The U.S. EPA and some states have specific rules about environmental audits as they apply to industry. The EPA audit policy is called Incentives for Self- Policing: Discovery, Disclosure, Correction and Prevention of

Violations. It does not cover all industries and provides substantial incentives for voluntary industry disclosure.

These incentives to the industries that come under the regulation occur only if certain conditions are met. The industrial facility must first voluntarily discover an infraction of environmental law. Many serious environmental impacts are allowed in facility permits so that any environmental impact may not necessarily be enough to require voluntarily disclosure. For purposes of sustainable development, this policy is inadequate because audits of all environmental impacts are necessary. After voluntary discovery, the industrial facility must quickly disclose the problem to the EPA. They must quickly correct the problem, and they must take steps to prevent the problem from recurring. In practice, industrial facility disclosures are preceded by discussions with the EPA. In these predisclosure discussions, a deal is developed that includes how and what must be disclosed, and how to do future audits. Many environmentalists feel that this approach does not adequately protect the environment or human health. EPA counters that without voluntary disclosure by regulated entities, there could be more impacts on the environment and human health because it would not be able to know all the problems of industrial nondisclosure.

Federal and state governments do not want to overregulate industries with environmental regulations so that they fail to make a profit. That is why they provide many incentives to industries to comply with environmental law. They do not seek to punish industries that break environmental laws unless there is a strong case that is repeated. Two components of a civil penalty are based on the severity of the violation. If the industrial facility voluntarily discloses, then the severity is reduced and so is the civil fine. The other component is the amount of economic benefit the industrial facility received as a result of its noncompliance. The EPA can take all or part of that economic benefit, but if it is voluntarily disclosed, it tends to be more lenient. Voluntary disclosure, usually via an environmental audit, can also prevent criminal prosecution for a violation of environmental law. Under all these circumstances, the industrial facility must voluntarily disclose in good faith. This is usually the result of an independent voluntary audit. Repeat violations, even if voluntarily disclosed through an environmental audit, are not eligible for any of the incentives. Repeat violations are closely related to the original violation within the past three years, or those that occurred as part of a pattern at multiple facilities owned or operated by the same entity. A new owner is eligible for a reduction of severity and incentive protection through voluntary disclosure, even if the facility had similar offenses prior to new ownership. There is considerable controversy about these policies and the EPA is considering eliminating them.

Many state environmental agencies have similar environmental audit, voluntary disclosure, and incentive policies. They are very protective of industry. Because the environmental audit is the basis for all these policies, many other stakeholders want them to be more transparent.

Most industries are either beneath regulatory thresholds or permitted to emit chemicals in the first place. If an environmental audit uncovers a problem, then even stricter scrutiny should be applied and the following environmental audits should be even more transparent. For purposes of sustainable policymaking and especially for application of the precautionary principle, this heightened transparency is more important.

The cumulative effect of environmental auditing is important for sustainable development. Environmental auditing should be done for all stakeholders and at any level that impacts the environment. Big and small industries, big and small communities, big and small organizations, and households all contribute to cumulative environmental impacts. Environmental auditing provides a needed baseline from which to judge environmental impacts.

LOCAL AND REGIONAL GOVERNMENTAL ACTIVITIES

Local and regional governmental activities can have large environmental impacts. How roads are built, where homes and businesses are allowed to grow, how water is used, where food is grown, and how air pollution is measured are all generally controlled by local and regional governments. Environmental policies in particular are affected at the implementation phase of policy development by local and regional governmental activities.

Local land use practices and policies from local governments do not communicate well with state environmental agencies. This is a weak link in U.S. environmental decision making. State environmental agencies issue permits to industry to emit chemicals into the land, air, and water. Societies with a high value on economic development through industrialization will place a priority on the speed of the transaction because time can erode profit. Industry is always complaining about the time it takes the government to process permit applications, renewals, and modifications while at the same time resisting even self-reporting of some environmental information. Local government land use decisions regarding industry is often in the same dynamic when industrial economic development is sought. Local land use policies and state environmental policies conveniently do not share much information about prospective new development. Because speed of transaction is important for economic development, time consuming and troublesome stakeholders that could stop a project are not included. Or, if the law requires some type of notice to the community, they are included in ways that are not meaningful. Industry will face a big challenge from the equity component of sustainability because this type of decision making is not inclusive and does not accurately represent past, present, or future environmental impacts.

References

Rogers, Heather. 2006. *Gone Tomorrow: The Hidden Life of Garbage.* New York: New Press.

Rosemarin, Arno et al. 2008. *Pathways for Sustainable Sanitation: Achieving Millennium Development Goals.* London, UK: IWA.

Waste Management

Garbage, hazardous wastes, and sewer and waste water all accumulate with human habitation. When fresh water and waste water meet, the water can become a vector for waterborne human diseases. Public health concerns often form the basis for waste management policies. Waste management can be public and private, personal and municipal.

In terms of sustainable development, the goal is to reduce wastes in order to reduce environmental degradation. In many countries, such as the United States, high levels of consumption go with high levels of waste production. Reusing some wastes as a product, recycling some wastes to decrease their environmental impact, and reducing wastes by increasing the efficiency of consumption are all part of that goal.

Technology is a waste creator on several levels. Even if the new technology has a lower environmental impact, decisions about what to with the old technology need to be considered for sustainability purposes. One area of rapidly improving efficiency is in the area of electronic microprocessors. With the technological changes of cell phones to mobile Internet devices, the need for power increases. The Intel Atom Processor is specifically designed for the increased energy needs and to decrease energy waste. Intel claims it is 10 times more efficient than its 2006 microprocessor. Increases in efficiency decrease power needed from other sources, and many of these other sources create carbon in the air, a greenhouse gas. Making silicon chips that are flexible may increase the applications of solar power generation. When printers became commonplace, the amount of paper and ink usage increased dramatically. Technology changes rapidly, and the waste streams from technological sources both increase and change as rapidly. In the United States, at least 3 million tons of computers end up in landfills annually. There they very slowly decompose and leach over 70 known hazardous chemicals into the environment. Environmental Assessment Tool (EPEAT) developed an electronic product in 2006 by the Green Electronics Council (see www.epeat.net). Its Web site gives access to this tool that evaluates the environmental impact of most computers. Part of its assessment is the chemical composition of a given computer, its energy efficiency, and whether the manufacturer will collect and recycle old machines. It evaluates computers and gives them gold, silver, or bronze awards based on the environmental impacts of the computers. According to the Green Electronics Council, 109 million EPEAT-approved computers were sold in 2007, in its first year of operation. This stopped about 124,000 metric tons of toxic materials from entering landfills and prevented about 3.3 million tons of greenhouse gas emissions in one year. Energy and technological waste streams pose a constant challenge to sustainable development because of the promise of decreased environmental impacts, the creation of waste from old technology, and their shifting energy sources.

Wastes are regulated more and more. The terminus for waste streams is often the ocean, called deep ocean dumping. Many land-based waste streams follow rivers out to the coastlines to create large "dead zones." Ocean-going ships often carry wastes or create wastes that they dump from bilges into the ocean and even into ports. In some parts of the Pacific and other oceans, there are large islands of plastic debris. These are accumulating in areas of the oceans that are still, and outside of the trade routes. They are very harmful to the marine life. Many nations deposit wastes in the deep ocean. Some scientists maintain that the type of waste determines whether deep ocean dumping is in fact environmentally degrading. Some mining wastes may be suited for deep ocean dumping more than land dumping because there is little life in parts of the deep ocean and it is so vast. There are few legally permitted deep ocean waste sites.

Watershed Activities

Activities in and around watersheds include all impacts on the watershed such as irrigation, recreation, waste management, and drinking water. A watershed is essentially an area of land where all the surface and ground water moves to the same location. In this way watersheds are the connecting force of most ecosystems. They are shaped continually by forces of nature. Watersheds include those forces of nature that affect them such as geology and air. Parts of watersheds, such as rivers, have served as political boundaries for centuries; however, the watershed is much greater than the river. As human population, waste, and environmental impacts have increased, the stress on watersheds has increased. This increases the stress and threat to natural ecosystems, which in turn increases the potential to create irreparable harm to systems of nature on which future life depends, a major tenet of sustainability.

The laws, rights, responsibilities, and ownership of water are increasingly controversial. Some predict that future wars will be based on water. The vice president of the World Bank said in 1995, "If the wars of this century were fought over oil, the wars of the next century will be fought over water." The interconnected nature of watersheds means that activities in one part of it could affect activities in another part of it, generally downstream. If one part is polluted by agricultural runoff, human waste, or industrial manufacturing processes, then other parts that do not create the problem may suffer the consequences.

Watersheds are important and many times unknown parts of the ecosystems. Underground water flows and aquifers are being pumped dry before their capacity is known. Increased rate of water use in one part of a given watershed will decrease the availability of water in other parts of the watershed. As water quantity decreases, the water quality also often decreases because of the higher concentration of contaminants. As climate changes occur because of global warming, the consequences for the world's watersheds are uncertain. This is a very important consideration for determining the ecological carrying capacity and appropriate environmental and sustainable development policy for developed and developing nations.

Watershed restoration is an important activity for sustainable development. It can challenge cultural values such as private property ownership, water as a sacred life force, and water allocation laws and contracts. Watershed restoration activities can be targeted for specific environmental restoration activities.

In Oregon, watershed restoration was developed in 1997 to restore native fish populations that were listed under the Endangered Species Act. The Oregon Plan for Salmon and Watersheds tries to restore watersheds for salmon runs by improving water quality. Salmon return to the place of their birth to spawn after years in the ocean. If the water is too warm, too turbid, or too polluted they will not spawn. The Oregon Plan for Watershed Restoration relies heavily on voluntary actions of citizens and industries. Rivers with salmon course through many rural areas with private property owners who are not required by law to restore rivers. The Oregon plan seeks to coordinate federal, state, tribal, and local government actions to assist private landowners. They do this by promoting public education and overall community awareness about the role of watersheds and salmon. The plan has a strong monitoring component for watershed conditions, water quality, and level of salmon recovery. This monitoring system is designed to document existing conditions, follow any trends, and evaluate the impact of the program. In some locations the monitoring system videotapes every fish that comes or goes through a river. The plan is overseen by an independent group of scientists charged with evaluating its effectiveness, recommending needed changes, and developing further research. The plan is run by the Oregon Water Enhancement Board and is funded through the state lottery and the sale of state license plates with a salmon on them.

The U.S. EPA has an extensive watershed policy. The economic base for most industries and communities is a clean and reliable watershed. The EPA recommends a strategy for prioritizing watershed resources. These priorities can be controversial and at cross-purposes, such as habitat preservation and agricultural irrigation. There are generally multiple users of a given watershed, and many take their use for granted. Some watershed management strategies are required by regulations, some are voluntary, and some mix both approaches. EPA's sustainability approach to watershed management recommends full cost pricing of water resources. Right now, many water uses are not even monitored. Some large water users are subsidized by government either directly or indirectly. Indirect water use subsidies occur when crops that use too much water for a given ecosystem, such as rice in a desert, are subsidized. Another important aspect of the EPA's approach to sustainability is the state of water infrastructure. Many of the pipes are old, and some leak. In water systems serving more than 100,000 people, about one-third of the pipes are between 40 and 80 years old. One-tenth of the pipes serving these populations are over 80 years old. Many of these systems have been added on to as populations increased. Waste treatment facilities and consolidated sewer systems with antiquated storm water overflows directly into rivers pose an increasing problem for systems of sustainability as populations increase and water infrastructure deteriorates. Water use is inefficient. Streamlining water use and improving infrastructure may aid in the preservation of exiting watersheds, but it may uncover serious problems.

International perspectives on watershed management vary widely. Some countries just accept a rapidly declining level of water quality. This can cause increases in waterborne public health threats. Other countries view water as sacred and a source of life. International sustainability initiatives examining ecosystems are focusing on discovering where the big watersheds are and how they function in a given ecosystem. From there they hope to work out the carrying capacity of that ecosystem for sustainable development.

Watershed activities are the leading edge of sustainable development and as such will face the most entrenched controversies first.

References

Conca, Ken. 2005. *Governing Water: Contentious Transnational Politics and Global Institution Building.* Cambridge, MA: MIT Press.

Kassim, Tarek A., and Kenneth J. Williamson. 2005. *Environmental Impact Assessment of Recycled Wastes on Surface and Ground Waters.* New York: Springer.

Land Use and Growth Management

Land use refers to the array of government controls over the use of land, such as zoning. Zoning divides the land into allowable uses such as residential, commercial, or industrial. Land that government owns is also use controlled, such as parks, mines, and monuments. The U.S. federal government is a large landowner, and state and local governments can also own land. Private citizens and sometimes foreign corporations can also own land, known as private property. In the United States, if the land use control takes away all the value of a privately owned parcel of land, it is called a "taking" of private property, and the government must compensate the owner for the fair market value of the land. Many local and state governments try to regulate land use without it being a taking of private property. Takings of private property are expensive and often unpopular with citizens, land speculators, and real estate development.

Growth management refers to the timed and sequential control of land uses. Municipalities (towns, villages, cities, and counties) try to provide the necessary infrastructure for the type of land use growth they want. Infrastructure refers to bridges, roads, pipes, dams, and communication and power distribution. A community can control the kind of growth it wants with the infrastructure it chooses to provide, often called capital improvements projects. Developers often want the municipality to provide the infrastructure they want to maximize their profits. Sometimes they claim to have a right to force the city to provide the infrastructure. The question of who bears the cost of infrastructure improvements is complicated. Neither party wants the other to fail, but both want the other to pay for the type of infrastructure that meets their goals. Sustainable development is seldom the goal of developers, and only occasionally the goal of cities.

Growth management of land uses has generally been devoid of serious environmental impact assessment until recently. Developers will mitigate environmental impacts only on site, at all. These environmental mitigations are poorly monitored and therefore poorly enforced. The actions claimed as environmental mitigation are often short-term efforts, such as laying straw down on a construction site to prevent erosion of soil. There is no requirement to try to lessen the impacts of environmental mitigation. So if the straw clogs the sewer and water system there is no requirement to mitigate that effect. Once the project is complete the developer may be hard to locate, and the financing organizations for the developer may claim to lack enforceable environmental requirements with the mitigation agreements. If, however, the project is financed with public funds, such as climate-neutral bonds, then some accountability may exist via the bondholders and the issuing municipality or state.

Growth management of land uses is fueled by population growth and the consumption of infrastructure by the current and new population. Population is increasing and so are the potential environmental impacts. Alternative technologies in energy may decrease the rate of environmental impacts. One of the issues in the forefront of sustainable development is the control of land use growth to not exceed the environmental carrying capacity of the systems of nature on which future life depends. Some U.S. cities have recently implemented the precautionary principle in their laws and regulations, but it is not known if they apply to growth management of land uses. If one of the goals of growth management were sustainable development, then aligning land uses with watershed practices could be a significant step for forward. *See also* **Volume 1, Chapter 5: Collaborative Decision-Making Processes.**

References

Benfield, F. Kaid et al. 2001. *Solving Sprawl: Models of Smart Growth in Communities across America*. Washington, DC: Island Press.

Helming, Katharina, Marta Perez-Soba, and Paul Tabbush. 2008. *Sustainability Impact Assessment of Land Use Changes.* New York: Springer.

Mander, Ulo, Hubert Wiggering, and Katharina Helming. 2007. *Multifunctional Land Use.* New York: Springer.

Platt, Rutherford H. 1996. *Land Use and Society: Geography, Law, and Public Policy*. Washington, DC: Island Press.

Porter, Douglas. 2002. *Making Smart Growth Work*. Washington, DC: Urban Land Institute.

Zovanayi, Gabor. 1998. *Growth Management for a Sustainable Future: Ecological Sustainability as the New Growth Management for the 21st Century*. Westport, CT: Greenwood Publishing.

Climate-Neutral Municipal Bonding

Some states and local governments have begun to connect their considerable financial powers to the cause of mitigating climate change. State and local governments can raise substantial sums of money by issuing tax-exempt bonds. These bonds are attractive to institutional investors. Some municipalities and states now require that projects funded with these tax-exempt bonds must not increase greenhouse gases. This is termed as climate neutrality.

State Lawsuits on Climate Change

State attorneys general for California, Connecticut, Iowa, New Jersey, New York, Rhode Island, Vermont, and Wisconsin sued the electrical utilities in their states. The wanted the courts to stop these plants from emitting so much greenhouse gas in the production of power from coal. Their theory in these lawsuits was that the defendants' emissions were a public nuisance. Nuisance is a common law doctrine from English common law. Courts refused to hear the cases, however. More lawsuits on climate change are likely to be filed as legislatures and the executive branch develop more public policy.

Reference

Connecticut et al. v. American Electric (SDNY, 2004) (1:04-CV-05669).

Gerrard, Michael B. 2007. *Global Climate Change and US Law*. Chicago: ABA Press.

Haroff, Kevin T. and Katherine Kirwan Moore. "Global Climate Change and the National Environmental Policy Act." *University of San Francisco Law Review* 42, no. 1 (2007):155–83.

Ohshita, Stephanie B. "The Scientific and International Context for Climate Change Initiatives." *University of San Francisco Law Review* 42, no. 1 (2007):1–37.

Precautionary Principle

When harm to the environment or human health will result from development, reasonable measures should be taken to prevent harm even if the scientific evidence is inconclusive. This effectively shifts the burden of proving risk of harm from a member of the public or a potential victim of harm to the proponent of activities that may cause harm.

This core principle from Agenda 21 has been applied by state and local governments to land use regulation. The National Association of County and City Health Officials adopted Resolution 03–02 on September 8, 2003, urging that the precautionary principle be applied to the activities of local health departments. This resolution illustrates the burden-shifting aspect of precaution.

WHEREAS, the precautionary principle states, When an activity raises threats of harm to human health or the environment, precautionary measures should be taken, even if some cause and effect relationships are not fully established scientifically. In this context, the proponent of an activity, rather than the public, should bear the burden of proof. The process of applying (this principle) must be open, informed and democratic and must include potentially affected parties. It must involve an examination of the full range of alternatives, including no action;

The resolution further suggests practical steps to implement the precautionary principle

- Integrate public health and land use planning,

- Full participation by affected communities in land use decisions that affect them, and

- Better training of public health officials to increase their capacity to understand and participate in land use planning decisions.

MUNICIPALITIES ADOPTING THE PRECAUTIONARY PRINCIPLE

The precautionary principle is widely used in Europe and increasingly so in Canada as well. It directs governmental agencies to restrict the use of products or compounds suspected of producing health risks. The precautionary principle is articulated in Europe as a measure requiring industry to prove that a product or compound is safe rather than showing it is not harmful. This may just seem like semantics, but the articulation has the effect of creating requirements that are more stringent and protects the health of the public.

A municipality is a unit of local government in the United States. Towns, villages, cities, and other political subdivisions of the state are legally called municipalities. It has power over much of land use and land development. In the United States, municipalities have not been environmentally sensitive but tend to be focused on economic development through real estate. The state can preempt the land use of the municipality and the federal government can preempt both the state and the municipality. Because the focus of municipal land use is generally economic development, it tends to make real estate transactions as efficient as possible to decrease transactional costs, such as those costs resulting from local regulations. Another factor is the highly held value of private property in the United States. This value also makes local land use regulation less involved in environmental issues.

Sustainability is directly concerned with the environmental impacts of past, present, and future local land use decisions. One of the ways that sustainability becomes involved with local land use decisions is through the application of the precautionary principle. The city and county of

San Francisco (Calif.) were the first to adopt this principle in June 2002. Two municipalities that formally adopted the principle are the City of Berkeley, California, and Lyndhurst, New Jersey.

THE SAN FRANCISCO EXPERIENCE: SUSTAINABILITY IN LOCAL U.S. GOVERNMENT

Both the city of San Francisco and the county of San Francisco adopted the precautionary principle first in the United States. These are wealthy municipalities on the West Coast with highly educated and diverse populations. They are not especially industrial in terms of economic development but do have high concentrations of population. They are nestled between the Pacific Ocean to the West and mountain ranges to the East. The city of San Francisco is backed up into a large bay.

The adoption of the first local law, called an ordinance, was preceded by intense political dispute. It is likely to be copied by other municipalities, and the two discussed here did adopt significant parts of their language. Sections of this new law are directly related to developing a policy of sustainability, and often do more than adopt the precautionary principle alone.

The local elected body, called the board of supervisors, made specific findings to presage the law. These findings were that every San Franciscan has an equal right to a healthy and safe environment. This requires that the air, water, earth, and food be of a sufficiently high standard that individuals and communities can live healthy, fulfilling, and dignified lives. The duty to enhance, protect, and preserve San Francisco's environment rests on the shoulders of government, residents, citizen groups, and businesses alike. These findings seem obvious to most people in most communities, but the local elected officials in San Francisco made them the basis for the adoption of the precautionary principle.

The San Francisco Board of Supervisors noted that historically, environmentally harmful activities have been stopped only after they have manifested extreme environmental degradation or exposed people to harm. They took special note of DDT, lead, and asbestos. In these cases regulatory action took place only after disaster had struck. The delay between first knowledge of harm and appropriate action to deal with it can be measured in human lives cut short. They considered themselves a leader in making choices based on the least environmentally harmful alternatives. They specifically rejected traditional assumptions about risk assessment and management. San Francisco already had other local laws that dealt with specific policies and that included the precautionary principle. These laws included the Integrated Pest Management Ordinance, the Resource Efficient Building Ordinance, the Healthy Air Ordinance, the Resource Conservation Ordinance, and the Environmentally Preferable Purchasing Ordinance.

San Francisco elected officials sought to integrate all these environmental laws into one policy approach under the broad rubric of the precautionary principle. They also sought to have a more just application of these environmental laws. *See also* Volume 3, Chapter 2. Their specific goal was to be sustainable by creating a sustainable San Francisco Bay area environment for present and future generations.

They included a focused discussion of the potential solutions, problems, and limitations of science in their findings that presaged the precautionary principle ordinance. They found that science and technology are creating new solutions to prevent or mitigate environmental problems. They also found that science is also creating new problems with unintended consequences. In their view, the precautionary principle is a policy to help promote environmentally healthy alternatives while preventing the negative and unintended consequences of new technologies. To do this, they use a form of alternatives assessment based on environmental impacts. This assessment is based on the best available science. The alternatives assessment examines a broad range of options considering short-term versus long-term effects or costs. It then evaluates and compares the adverse or potentially adverse effects of each alternative, assessing options with fewest potential hazards. This includes the option of doing nothing.

Another key principle of sustainability, transparent environmental transactions, is applied to the alternatives assessment under the precautionary principle ordinance. The alternatives assessment is a public process because, in the view of local elected leaders, the public bears the ecological and health consequences of environmental decisions. San Francisco's local leaders seek to fully engage the community in meaningful ways to the alternatives assessment process because they found that the final decision is more robust by broadly based community participation in alternatives assessment. In this process a full range of alternatives are considered based on input from diverse individuals and groups. The community should be able to determine the range of alternatives examined and suggest specific reasonable alternatives, as well as their short- and long-term benefits and drawbacks. This assumes the community has the capacity and time to do so. Citizens are assumed to be equal partners in decisions that affect their environment.

The local elected leaders of San Francisco based their findings for the new precautionary principle on the goals of a future where the city's power is generated from renewable sources, when all waste is recycled, when vehicles produce only potable water as emissions, when the San Francisco Bay is free from toxins, and the oceans are free from pollutants. These goals are specifically founded in a hope for sustainability. They intend to use the precautionary principle as a means to help attain these goals and to evaluate future laws and policies in major areas as transportation, construction, land use, planning, water, energy, health care, recreation, purchasing, and public expenditure.

The San Francisco Board of Supervisors was also cognizant of the human behavior changes that will be necessary to live sustainably. They found that the precautionary principle as a policy approach will help San Francisco speed this process of human behavioral change by moving beyond finding solutions for environmental degradation to preventing environmental harm.

To implement the application of the precautionary principle, there must first be reasonable grounds for concern. This will be a matter of interpretation by city planners and managers. Where there are reasonable grounds for concern, the precautionary policy approach will be to reduce environmental harm by starting a process to select the least potential environmental threat. This process is made of principles to guide the implementation and application of the precautionary principle, and these are some of the essential elements of sustainability.

The first is anticipatory actions to prevent environmental or human harm. The stakeholders of government, business, community groups, and the general public share this responsibility and duty to engage in anticipatory actions, but these actions are as yet ill defined. If a particular stakeholder does know that an action will cause harm, then there is a duty to prevent it.

This duty is based on knowledge, and another guiding principle of the precautionary principle law is an expansive right-to-know provision. The community has a right to know complete and accurate information on potential human health and environmental impacts associated with the selection of products, services, operations, or plans. Unlike many other right-to-know laws, the burden to supply this information lies with the proponent of a particular action, not with the community or government. There is also a strong duty to do full cost accounting in the assessment of alternatives to a proposed action. This is very extensive and includes raw materials, manufacturing, transportation, use, cleanup, eventual disposal, and health costs, even if such costs are not reflected in the initial price. Short- and long-term time thresholds should be considered when making decisions. Although not specifically stated in the precautionary principle ordinance, these could include the cumulative impacts of a particular alternative. At the end of three years, the local environmental agency, called here the Commission on the Environment, will evaluate the effectiveness of the precautionary principle ordinance and submit a report.

The San Francisco ordinance incorporating the precautionary principle as an overarching policy to local land, health, and environmental decisions is a new and foundational development in sustainable development. There are many questions about its costs, benefits, effectiveness, and interaction with other levels of government in the United States. Some view it as the U.S. catching up with European municipalities and their approaches. Others view it as economically unworkable, overburdensome to some stakeholders such as real estate development, and beyond the capacity of local government and citizens.

THE BERKELEY PRECAUTIONARY PRINCIPLE AS LAW

The municipality, or city, of Berkeley California specifically adopted the precautionary principle in its land use laws, called ordinances. Berkeley is a city near the city of San Francisco. Like San Francisco, it has a highly educated, relatively wealthy, diverse population with some industry. A primary employer is the University of California at Berkeley, a leading U.S. research center. The specific purpose in the city of Berkeley in adopting the precautionary principle as an overarching policy is to protect the health, safety, and general welfare of the city. The law does this by decreasing health risks, improving the air quality, protecting the quality of the ground and surface water, decreasing resource consumption, and decreasing the city's impacts on global climate change. It intends to implement its precautionary principle law in phases.

The city defines its precautionary principle policy to mean where threats of serious or irreversible damage to people or nature exist, the lack of full scientific certainty about the case and effect shall not be viewed as a sufficient reason for the city to postpone measures to prevent the degradation of the environment or to protect human health. Furthermore, any gaps in the scientific data discovered by the examination of alternatives will serve as guides to future research on the issue. These scientific gaps will not prevent the city from taking actions protective of the environment. As new scientific information becomes available, the city can review these decisions and make any necessary changes.

The city of Berkeley intends to apply the precautionary principle under a set of guidelines. It will use anticipatory actions to prevent harm to the environment or to human health. It specifically states that government, business, community groups and the public all share this responsibility to engage in anticipatory actions to prevent such harms. It is also guided by an extensive right-to-know requirement. The community has a right to know complete and accurate information on the real and potential health and environmental impacts associated with the selection of products, services, operations, or plans. This is one of the most extensive right-to-know laws in the United States. The city is also guided in its application of the precautionary principle by a requirement that all alternatives be fully assessed, and that the alternative with the least potential impact on health and the environment be selected. It must consider the impact of doing nothing as one of the alternatives. It is also guided by a full cost comparison. It needs to consider long- and short-term costs of any product, including an evaluation of the significant costs during the lifetime of the product. This includes raw materials, manufacturing and production, transportation, use, cleanup, acquisition, warranties, operation, maintenance, disposal costs, and long- and short-term environmental and health impacts. These costs are compared to any available alternatives.

The city's application of the precautionary principle is also guided by a participatory decision process that is transparent and uses the best available information. The city's current sustainable practices are incorporated into the new precautionary principle law. The city is to purchase products or services that reduce waste and toxics, prevent pollution, contain recycled content, save energy and water, follow green building practices, use sustainable landscape management techniques, conserve forests, and encourage agricultural biobased products. The city uses redwood that is certified as sustainably harvested. It will decide how the precautionary principle applies to future actions, based primarily on whether the city manager considers it feasible at the time. The new precautionary principle law also requires that an annual report on implementing actions be produced and available to the public.

The precautionary principle as urban policy in Berkeley is slightly more specific than the San Francisco law. It is reviewed annually, whereas the San Francisco precautionary principle law is reviewed every three years. Because the Berkeley policy is more specific and reviewed more, it may be a better research basis for analysts of sustainable development in U.S. local government. It is likely that state and federal levels of government will examine these laws closely. Some fear that state and federal government agencies will use their preemptive power to usurp these local initiatives before analyses of their effectiveness can take place. Others feel that the momentum of international sustainability movements will prevent the use of the preemptive power of higher levels of government.

Neither of these precautionary principle ordinances were formed in a particular controversy. They were both formed around a hopeful goal of sustainability for the future. Other older and more industrial cities also want to shape sustainable urban policies. These precautionary principles can be formed in the heat of environmental controversy.

LYNDHURST, NEW JERSEY

Lyndhurst, New Jersey, is a municipality in the United States with heavy industry and environmental concerns about cancer clusters. After San Francisco and San Francisco County, it was the second city to adopt the precautionary principle into law. In an effort to protect its citizens from the environmental impacts of heavy industry, the mayor and his staff developed a precautionary principle following international definitions and most of the San Francisco model.

The municipality of Lyndhurst, known as a "township" in New Jersey, adopted the precautionary principle as the policy of Lyndhurst. The city basically adopted the international statement of the precautionary principle, which is that when an activity raises threats of harm to human health or the environment, precautionary measures should be taken, even if some cause-and-effect relationships are not fully established scientifically. Although this is a broad principle, Lyndhurst narrows its application through its implementation. All its officers, employees, boards, commissions, departments, and agencies are required to implement the precautionary principle when conducting town business.

Lyndhurst further anticipates implementing the precautionary principle by developing new laws to create a healthier environment. It specifically wants to be a model of sustainable development by creating and maintaining a viable and healthy environment for both current and future generations. Its version of implementation of the precautionary principle is as both a policy tool and overall philosophy to advance environmentally healthy alternatives and remove negative and unintended consequences of modern and future technologies.

Another way Lyndhurst seeks to implement the precautionary principle is through an aspect of environmental justice. It wants to provide every resident with an equal right to a protected and safe environment and avoid disproportionate impacts. It will seek to find the least environmentally harmful alternative in every decision it makes. To use the precautionary principle for a healthy environment, Lyndhurst takes a strong ecological perspective of seeking a high standard of safety in the land, air, and water. It specifically wants to prevent environmental degradation and human health risks before they occur, not after they harm people or the environment. Whenever Lyndhurst is faced with a decision, it will analyze and assess all the alternatives. It intends to assess a wide range of alternatives and consider short- and long-range impacts.

Under this type of analysis it may get to cumulative impact analyses. As Lyndhurst is an older industrial township, these impacts could be significant. The policy of the precautionary principle is to compare and contrast the adverse and potentially adverse effects of each option, specifically noting alternatives with the fewest hazards. There are many environmental decisions with potentially adverse impacts that can include those not scientifically proven or disproved. Lyndhurst is sensitive to potentially hazardous activities and will ask if the activity is even necessary under the precautionary principle. It will try to measure how much harm it can avoid in its assessment of alternatives under the precautionary principle policy. This approach is very unusual and a foundational development in U.S. sustainability. It places the burden of proving the safety of an activity directly on the proponent, and not on those opposing such activities such as the community or local government. If there is a threat of irreversible damage to the community or the environment, its policy explicitly does not accept the lack of full scientific certainty about the causes and effects as a legitimate reason to delay governmental intervention that protects the environment or the community. The township will revisit decisions as new scientific information becomes available. This may have the effect of increasing scientific knowledge and community-based monitoring of environmental decisions, both of which are necessary for sustainable development.

Like the Berkeley and San Francisco laws, the Lyndhurst precautionary policy has a set of guiding principles. First, there must be some reasonable basis for concern. Lyndhurst does have clusters of cancer and a history of heavy industry. Whether this alone is a reasonable basis for the application of the precautionary principle is unknown but would depend on the type of project or decision proposed. Nonetheless, if there is a reasonable basis for concern, then the precautionary policy creates a duty of anticipatory action to reduce harm. The law creates this duty in government, community, business, and the general public.

To anticipate environmental or community harm, it must first be known. The next guiding principle is a powerful right-to-know law that places the burden of creating the information on the project proponent. It specifically states that the community has the right to know complete and accurate information on the actual and potential health impacts of the alternatives under assessment.

And alternatives are required to be assessed, including the no action alternative. An obligation is created to select the alternative with the least harmful impact. Alternatives are also to be evaluated under the principles of full cost accounting. This is a comprehensive cost accounting measure that evaluates all the costs including raw materials, manufacturing, transportation, cleanup, waste disposal, health impacts, and any others. It examines the entire lifecycle for the product or service, from cradle to grave. The entire process is to be transparent in that the community and public at large are to have meaningful involvement in finding and selecting alternatives under assessment.

Lyndhurst's precautionary principle policy is based expressly on the preservation of the environment. It creates a duty to enhance, protect, and preserve the environment on every citizen, business, nonprofit, and branch of township government. Under this law, the township wants to generate energy from renewable resources, recycle waste, and prevent pollution and toxins from entering their ecosystem. It views this law as providing the vehicle to develop future laws and policies that protect the sustainability of the ecosystem in the areas of land use, urban planning, water, energy, health care, recreation, transportation, government purchasing, and all public expenditures (such as education).

Lyndhurst developed its precautionary principle as law and policy in the context of a heated history of environmental and public health controversy. Court cases and scientists could not resolve this longstanding, simmering, adversarial standoff. Many municipalities are in this position in the United States and around the world. Economic development, capitalism, private property, and scientific postulations of cause and effect are now being directly challenged by communities and their democratically elected leaders. The result is a policy that goes beyond environmental preservation and conservation, beyond minimal protection of the public health, and that rejects current risk assessment models as inadequate. The result is a growing urban policy that is explicitly ecosystem-based, dynamically inclusive of community in meaningful ways, and that reorients local decision making toward precaution.

As these policies become implemented, they will create more controversy. Whether they are adopted by other municipalities, states, or the federal government, or are contested in those arenas, the controversy will continue. They are new, untested, costly, and may reveal unsavory truths about our past actions toward nature.

References

Myers, Nancy J., and Carolyn Raffensperger, eds. 2005. *Precautionary Tools for Reshaping Environmental Policy.* Cambridge, MA: MIT Press.

Whiteside, Kerry H. 2006. *Precautionary Politics: Principle and Practice in Confronting Environmental Risk.* Cambridge, MA: MIT Press.

CHAPTER 4

Controversies

Controversies are part of any major change in ways of thinking about the environment. They are disputes between groups or individuals. In terms of sustainability, controversies are mainly about environmental issues, and then about the policies necessary to mitigate impacts on natural systems. There are also major clusters of controversies around the business and economic impacts of sustainability and about the equity and fairness aspects of sustainable development. There will be many controversies surrounding sustainability.

As issues of environmental controversy become issues of sustainability, social issues will merge with environmental issues. Stakeholders are generally defined as those groups or individuals that have a stake in, or are affected by, the outcome of an environmental decision. Traditionally, they were first environmentalists and industry. As government intervention grew to mitigate some of the environmental impacts of industrial expansion, the government became another stakeholder in many environmental decision-making processes. Under social policies of sustainability, even those environmentally based, stakeholder groups increase. State, local, and tribal governments are now developing environmental policies. Industries differentiate their stakeholders as big or small industries and as green industry versus resource extraction industry. Environmental groups are challenged directly and indirectly by environmental justice groups.

The range of environmental decisions that will be examined under emerging sustainability policies will expand into many past and present environmental decisions. This includes many decisions that were not considered environmental decisions in the past such as decisions that ignored cumulative environmental impacts, ecological impacts, environmental impacts on cities, land use decisions, and environmental impacts on public health. Risk assessment and monitoring are likely to expand. This may change the role of many social institutions such as higher education by requiring a greater emphasis on environmental assessment and monitoring in faculty and student development.

Controversies are always shrouded in many questions. A big question is how to handle these controversies around environmentally based sustainability. Which stakeholders will get a seat at the decision-making table? What will be the theory of punishment for those who violate the principles of sustainability and create environmental degradation? What will be the role of science? What will be the role of the community? What if the environmental area is one where there is no human community? Who pays the price for these new processes and policies? Do economic and political systems that have rewarded us with freedom and a higher and longer quality of life have to change? Is such a change possible? These questions as yet have no ready answers.

Reference

Collin, Robert W. 2008. *Battleground: Environment.* Westport, CT: Greenwood Press

AGRICULTURE

Agriculture has been described as the systematic disruption of ecological systems to meet human needs. Some have questioned whether that can be done without degrading the environment. Agribusiness, which practices agriculture and animal husbandry on an industrial scale, has provided food quantities that have established food security in developed nations and surpluses. But the effects of this scale of agriculture on farming families, their communities, and the webs of life have been high. Among these consequences are increased pollution of our land air and water, including water pollution, run-off, soil loss, salinization, vast increases of toxins in our water and food, and vast increases in animal waste. Agriculture's role in history was to produce food for a growing population. Agriculture's role under sustainable development is to treat natural systems sustainably. Sustainable agriculture is an important component of sustainable development.

Agriculture and Soil

The soil is an integral component of agriculture. Human civilizations with good soil have lasted longer than those without good soils. Good soil, as a natural system, is created slowly but can be easily destroyed. Depending on climate, geology, and slope, soil is "created" at the rate of anywhere between .00067 and .00315 inches a year. This means it takes about 600 to 1,500 years to form 1 inch of soil. In areas where the topsoil is thin, agricultural practices can irreparably damage this natural system in a relatively short time of 500 years.

References

Davidson, John H. 2002. "Agriculture." In *Stumbling Toward Sustainability.* John C. Dernbach, ed., 351. Washington, DC: Environmental Law Institute.

Manjit, Kang S., 2007. *Agricultural and Environmental Sustainability: Considerations for the Future.* Boca Raton, FL: CRC Press.

Montgomery, David R. 2007. *Dirt: The Erosion of Civilizations.* Berkeley: University of California Press.

Agricultural Technology and Soil

Along with civilization, technology increased the efficiency of agriculture. One of the primary instruments of agricultural technology is the plow. Most farmers in the world use it, mainly for sowing crops. The reason for plowing is to bury weeds with last season's crop remnants and whatever soil amendments added in the course of the year. Plowing also opens up the soil, allowing exposure to air. This can accelerate some decomposition processes that release nutrients like nitrogen in the soil. Depending on climate and local conditions, plowing dark soil can increase the temperature in the soil. In northern and some southern latitudes, early warming of the soil can expand the growing season and range of plants under cultivation. Early soil warming can also accelerate germination times for some plants.

Plowing the soil, however, also allows for greater exposure to wind and rain. This can cause loss of valuable topsoils and erosion. Plowing is considered by many to be the main cause of degradation of agricultural soils. It also increases the vectors of ecosystem exposure from pesticides, fertilizers, soil sediment, and other wastes. As climate change expands desertification in highly populous areas around the equator, the loss of food production from agricultural land degradation in the context of still increasing populations pushes policies of agricultural sustainability.

Plows developed over history, from sticks used to plant seeds, to the use of animals to pull stick-made plows, to iron-tipped plows. Later the mold board plow was developed. This plow had a curved blade that turned the soil over on itself instead of just tearing open a furrow to plant seeds. In 1837, John Deere invented the sod buster, a tough steel moldboard plow specifically designed to break up U.S. tall grass prairie soil. This soil was very dense with roots, and often gummy. When the tractor was introduced to agriculture, many plows could be pulled behind it, as well as other soil treatment technology like discs and manure spreaders. Treatment of the soil is called tillage, and it can be an expensive cost of doing business for farmers and create environmental impacts such as loss of topsoil. In the late 1940s and early 1950s, a series of herbicides were introduced to agriculture. These had the effect of reducing tillage, but some also had environmentally degrading and persistent impacts on the environment.

Plows and plowing the U.S. prairies extensively allowed for soil to be blown away in storms, causing crop failure and the Dust Bowl. It also had the effect of jump-starting U.S. land conservation practices. The Dust Bowl affected the United States from 1931 to 1939. This spurred the development of the U.S. Soil Conservation movement. The U.S. government formed the Soil Conservation Service, which is now the Natural Resources Conservation Service.

Natural Resources Conservation Service

The Natural Resources Conservation Service developed the Ecosystem Sustainability Framework for County Analysis. This framework is based on an agricultural perspective.

The Ecosystem Sustainability Framework for County Analysis project examines economic and environmental areas of agricultural counties. The model assumes an agricultural system is sustainable if it meets indicator threshold levels of sustainability. One observation that is part of this framework is that natural systems can be compared for individual resource concerns.

Sustainable Alternatives to Conventional Agriculture: No-Till Farming

Conventional plowing leaves the soil exposed to wind and rain, is prone to erosion, and spreads impacts to a large part of most ecosystems via agricultural runoff of pesticides and fertilizers, and sometimes topsoil. No-till farming promotes minimal soil disruption. In this approach, farmers leave crop residue on the field. They sow seeds with a specialized machine that makes a small slit for the seed. Leaving the soil undisturbed promotes soil stability and growth, reduces water needs, prevents runoff of pesticides and fertilizers, and decreases the costs of tillage in areas other than sowing seeds. In can also increase biodiversity for many species from worms to birds. A key point of no-till farming for sustainable development is that it can sequester carbon and decrease emissions of greenhouse gases, thereby reducing the impact on global warming. Most crops take carbon dioxide out of the air and produce oxygen in a process of photosynthesis. The carbon dioxide remains in the roots and crop remnants. Between 1982 and 2003, according to the U.S. Dept of Agricultures National Resources Inventory, soil erosion of U.S. cropland decreased 43 percent. Many claim this is a result of increasing adoption of no-till practices.

These practices, however, can lower crop yields and increase herbicide use. The increased moisture retention can introduce new diseases. In some instances, no-till farming may require more nitrogen at first. Without the soil-warming effects of plowing, no-till farming may also have slower and later germination at northern and southern latitudes. Some water-intensive crops, such as rice, are not yet appropriate for no-till approaches.

Agriculture, Soils, and Fuel

Should agriculture grow fuel? The use of corn-based fuels has greatly increased the amount of cropland devoted to corn. If biofuel crops are a wave of the future, soil sustainability will be a key factor in mitigating the environmental impacts from such a large increase in production demand. Corn, for example, is the basis of many currently new biofuels.

When growing, corn does not get its nitrogen from the air, as many other crops do. As a result, the nitrogen loading necessary for soils to sustainably raise corn may be larger.

Soil and Waste Capacity

Part of the ecological footprint of all human habitation is waste. Waste increases as population and patterns of conspicuous consumption increase. The type and amount of soil are also important to the amount and type of waste it can handle. The capacity of landfills is an important and often controversial issue when protecting soil-based natural systems from environmental impacts. As landfills reach capacity, they can erode natural systems; however, waste must go somewhere. With the types and amounts of waste increasing, waste that is not recycled or reused is placed in waste transfer stations. These are temporary storage places, usually located along major waste transit routes. Typical waste transit routes are major roads, rail lines, and both inland and ocean shipping routes.

Because soil is a natural system, the impact of waste on it is a great concern for sustainability advocates. Soil is necessary for agriculture, but as landfills reach capacity, new landfill pressure may increase on agricultural lands. Two confounding issues are that agricultural industries can also produce large amounts of waste that degrade the environment, and sometimes "waste" from urban areas and the agricultural business can be reused either in the production process or in other industries or natural processes. Many cultures in the world use human waste as fertilizer and animal waste as fuel and building materials. Organic wastes can be remediated by healthy soil. Chemical wastes can cause environmental degradation and may have to work their way through ecosystems.

References

Montgomery, David R. 2007. *Dirt: The Erosion of Civilizations.* San Francisco: University of California Press.

Rogers, Heather. 2006. *Gone Tomorrow: The Hidden Life of Garbage.* New York: New Press

Spellman, Frank R. 2007. *Environmental Management of Concentrated Animal Feeding Operations.* Boca Raton, FL: CRC Press.

Irrigation and Drainage

Runoff nutrients contribute to water pollution and acidification. Insecticides, rodenticides, and fungicides add to crop yields but pollute water and can have harmful effects on farm workers, their families, and the ultimate consumer.

Farming is still among the most hazardous ways to earn a living, with conditions largely outside the standards of protection given to other workers. Farm workers and farm families often have no health care provision outside of emergencies. Industrialized agriculture has also eliminated support services that provided the basis for local economies, leaving farm communities struggling to survive.

Recognizing some of these devastating environmental and community conditions, some governments provided subsidies to try to encourage healthy and environmentally suitable agricultural practices such as soil conservation, crop rotation, and family farming. Over time, many subsidies were given to support crops such as tobacco and unsuitable land practices such as irrigated crops in artificial settings, which can contribute to the salinization of soil. Recipients are often agribusinesses, not family farm operations. Recipients often view subsidies as a form of entitlement and resist changes in policy regarding subsidies whether these changes are the result of free trade agreements or other policies.

Agricultural trends indicate that more farming operations will operate in a factory-type setting. This includes bringing more animals indoors for processing into food products. The scale of these factory-type operations has a much heavier ecological footprint on the place and the people who live and work there. Sustainable agriculture would not waste or pollute its ecosystems, will support its communities, and can maintain itself dynamically over time by adapting to changing circumstances.

Urban Farming

Urban farming refers to farming, or growing food, in urban areas. It is generally difficult to farm in urban areas because of the lack of space, knowledge of farming, and security for the crops in the growing area. It is difficult to farm in urban areas in a way that makes a profit so that most urban farming is for sustenance.

Urban farming is making a comeback as traditional farming methods require transportation to get from field to market. Transportation costs add to the price of food. The lack of land is a challenge for urban farmers. There are also new farming technologies, such as vertical farming, that encourage urban farming. Some communities are beginning to provide land for community gardens. Security for the growing crops is important. People need the incentive to harvest their crops before they are stolen.

Sustainability is served by urban farming because it decreases the environmental impacts of food production, storage, transportation, and packaging. The necessities of farming, soil, water, and sunlight are often in shorter supply in urban areas. For this reason, cities pursuing sustainable development need to plan around their urban farming needs. *See also* Volume 1, Chapter 5: Collaborative Decision-Making Processes.

Reference
Nordahl, Darrin. 2009. *Public Produce: The New Urban Agriculture.* Washington, DC: Island Press.

Organic Farming

Organic farming is a method of agricultural production that does not use chemical pesticides or fertilizers. Chemical farming methods expand the range of crops that can grow in a given area, expand the range of seasons to grow those crops, and increase the productivity per acre of land. As a result, they also help increase the profits of agricultural businesses and farmers. The increasing use of pesticides and fertilizers may negatively affect the environment when the se products run off into the watershed. For example, one factor in the destruction of the Great Barrier Reef off Australia is thought to be the high nitrogen and phosphorus content of the water runoff from agricultural lands. They encourage algae blooms that can increase the water temperature and attract starfish that destroy living coral reefs. Chemical farming or conventional farming tends to use large-scale approaches in a way that is not site specific, ignores ecological balances, and poses threats to biodiversity. Organic farming is a better fit with the carrying capacity of the ecosystem.

Organic farming uses natural methods of growing crops, especially in the soil. It is generally more labor intensive. Part of the movement of organic farming was the idea of returning waste to produce soil that would in turn produce food. This cycle would, in theory, keep soil quality high even as population increased in the future. By preserving systems of nature on which future life depends, organic farming was tied to modern concepts of sustainability. One modern version of organic farming is biodynamic agriculture. In this system the soil and planets are life forces that work together to make the farm an integrated, holistic unit. Another modern version of organic farming is organic gardening. Organic farming approaches are applied to gardening approaches that raise plants for recreation, aesthetics, or landscaping.

In the United States, the National Organic Program enacted a definition of organic farming in 2002. The definition of organic farming is "A production system that is managed with the Organic Food Production Act . . . to respond to site specific conditions by integrating cultural, biological, and mechanical practices that foster cycling of resources, promote ecological balance, and conserve biodiversity" (Volume 7 of the Code of Federal Regulations, Part 205). Critics have charged that this definition increases the commercialization of organic farming and erodes its small farm, rural character.

Organic farming respects the systems on which future life depends, and as such is part of sustainable development. However, it does not produce high profits, does require labor, and may limit agricultural productivity in some ecosystems.

References

Duram, Leslie, 2005. *Good Growing: Why Organic Farming Works.* Lincoln: University of Nebraska Press.

Guthman, Julie. 2004. *Agrarian Dreams: The Paradox of Organic Farming in California.* San Francisco: University of California Press.

Kral, David M., ed. 1984. *Organic Farming: Current Technology and Its Role in Sustainable Agriculture.* Madison, WI: American Society of Agronomy.

Conflict of Klamath River

The Klamath River runs for 263 miles in southern Oregon. It has many uses, from potato farming, to recreational fishing, boating, fishing rights of indigenous people, entrepreneurial activities of many people, and municipal water uses. All these stakeholders have been promised water from federal agencies. When there is enough water from rains and mountain snowmelt, there is little conflict. When there is not enough water, however, conflicts arise among water users. These conflicts can become violent because many water users depend on the water for their livelihood. The conflict becomes environmentally entangled with the rights of endangered species such as the sucker fish. The fish needs a certain water quality and quantity to survive. In 2002, there was a large fish die-off and measures were taken to preserve their environment. Some of these measures prevented agricultural users from using water necessary for their crops. One of the crops grown is potatoes, which consume a large amount of water, and some family-held farms were denied water, jeopardizing their livelihoods.

Scientific studies following the water temperature as it relates to the survival of endangered fish like the suckers continue. Some of the tribes protest the lack of water for salmon, for which they often have treaty rights to fish. The population of people in the municipal areas is increasing, which causes an increase in water use. Farmers, large and small, occasionally defy court orders and agency rulings to get water. Social tensions among indigenous peoples, local farmers, and grassroots environmentalists spill over into school buses, home life, and any type of collaborative approach.

As global warming continues, drought may increase and without adequate water conservation, planning, and watershed management, human conflicts will increase. The case of Klamath River is an important study of how a nonsustainable approach to watershed management only gets worse.

Reference

Klamath Basin Index. www.andykerr.net/Klamath Basin/KlamathBasinPT.htm.

National Academies Press. 2004. *Endangered and Threatened Fishes in the Klamath River Basin: Causes of Decline and Strategies for Recovery.* Washington: DC: National Academies Press.

Permaculture

Permaculture is a system for design within ecosystem capacities. Its early emphasis was on growing food within the capacity of a given ecosystem. It has since expanded to including building design, land use designs, and landscape restoration. The relationships of permaculture to ecosystems tie it directly to sustainable development.

Permaculture goes beyond organic farming because it looks at food in the context of the complete ecosystem and its natural range of biodiversity. By making the complete ecosystem more diverse and healthy, food productivity also increases. Food includes plants and animals. Birds are considered important parts of the ecosystem for their pest control, waste, seed propagation, and use as food. Chickens are often considered a part of permaculture if they are free ranging.

An important part of any ecosystem is the watershed. Most development uses impermeable surfaces that greatly increase runoff of water instead of letting it filter through the soil. The increased runoff can cause soil erosion as well as remove valuable soil nutrients. In nature, ecosystems absorb water through soil. Water in the soil also helps the soil become more productive by encouraging biological processes that can break down wastes into nutrients for plants. Ecosystems have different types of soil and different amounts of rain. Some ecosystems have soil that can retain more water than other parts where it may be rocky. Some ecosystems in dry areas may need to retain more water in order to grow plants and maintain soil-based biological processes. Permaculture will work with these carrying capacities of the ecosystem.

Permaculture is experiencing a rapid period of growth as sustainable development becomes a priority in many societies. For permaculture to be successful, an accurate and complete knowledge of the ecosystem is necessary, including human impacts.

Part of the recent growth of permaculture includes reviving environmental knowledge bases of traditional and indigenous societies, collecting and preserving older seed stocks and animals, developing urban planning principles, and building sustainable and intentional communities. Critics do not think that permaculture has proven itself to be productive enough to provide adequate food for a growing population. Some critics charge that permaculturalists bring in exotic and nonindigenous species into ecosystems to justify human food production.

References

Allen, Jenny. 2002. *Smart Permaculture Design*. Sydney, Australia: New Holland Publishers.

Tow, Philip, Ian Cooper, and Ian Partridge. 2009. *Rainfed Farming Systems.* New York: Springer.

Vertical Farming

Vertical farming is a way to grow food in an increasingly urbanized world. Not only is the world becoming more urbanized, the population will increase dramatically as will the need for food. Food security will become a major issue. Traditional farming methods will not be able to keep pace. Some estimate that more than 80 percent of the land available for traditional farming methods is already in use. If organic and permacultural methods are used, there may be decreased productivity, further decreasing the ability of traditional agriculture to keep up with food needs.

Farming vertically allows for greater productivity near the source of consumption on a year round basis. It may also decrease the pressure on already environmentally degraded ecosystems and be flexible enough to handle climate changes caused by global warming. The amount of uncertainty in traditional agricultural food production because of climate changes is very great. This increases the amount of risk to food security.

There is some dispute about how much increased productivity vertical farming offers. It depends on the crop and management systems. Generally one indoor acre is equal to about 4 to 30 acres outdoors. There is also less crop loss because of weather, and pests are more easily controlled. It may make organic farming approaches easier. Also, water is more efficiently contained. Some vertical farming advocates claim that it can create energy through the methane produced by composting organic matter. In terms of uses of nonrenewable energy, vertical farming produces food near the source of consumption, eliminating energy use in the distribution of food as well as energy use in the production of the food.

Vertical farming offers the potential to increase food productivity near population centers. Food security decrease is the first human sign that natural systems on which present and future life depends are eroding. Vertical farming will protect and preserve ecosystems to the extent that traditional farming methods become constrained by the use of vertical farming. The design of urban systems will have to change to accommodate vertical agricultural production. Sunlight, higher building stress levels, waste treatment, and waste transfer stations all have to be incorporated.

Reference

Bailey, Gilbert Ellis. 2008. *Vertical Farming.* Glacier National Park, MT: Kessinger Publishing.

Biodiversity and Monoculture: Food Security

Each ecosystem presents dynamic conditions for successful agriculture and for failure. Nature itself presents a broad range of plants uniquely adapted to these changing conditions. Humans have selected and manipulated the cultivation of these plants through agriculture. Agriculture cultivates plants for food, fiber, and fodder. Human selection of plants has narrowed based on market considerations including time to maturity, spoilage in transportation, labor handling, consumer preferences, and advertising. In addition, science now allows humans to genetically alter plants based on these preferences (genetically modified organisms). The power to select plants to cultivate has increased the amount of food available to human populations, but this choice has also eliminated the cultivation of many native plants.

Many native plants fit better into their present ecosystem because they can survive the diseases and climatic conditions present there. They may require less water and fewer pesticides, and may tolerate the presence of other plants. They can be grown closer to the point of consumption, requiring less transportation and decreasing environmental impacts. Native plants may also fill niches in the local ecosystem for life other than humans.

Nations that have monetary debt to organizations like the World Bank may have to grow nonnative plants for cash to pay these debts. These crops are known as "cash crops." As these nations grow dependent on money for modern items like cars, televisions, and refrigerators, their cash debt increases and their agricultural production focuses more on cash crops. Eventually they must import food and pay for imported food, which increases their debt and the production of cash crops that use nonnative plants. Many island nations rely heavily on imports for a modern lifestyle and have forsaken native crops for cash crops. The Western lifestyle in many contexts increases the environmental degradation of sensitive island ecosystems and vulnerability to famine by overreliance on nonnative cash crops. ***See also* Volume 1, Chapter 2: Biodiversity; Volume 1, Chapter 4: Agriculture.**

References

Bacon, Christopher M., et al. eds. 2008. *Confronting the Coffee Crisis: Fair Trade, Sustainable Livelihoods and Ecosystems in Mexico and Central America.* Cambridge, MA: MIT Press.

Lyson, Thomas, G. W. Steverson, and Rick Welsh, eds. 2008. *Food and the Mid-Level Farm: Renewing an Agriculture in the Middle.* Cambridge, MA: MIT Press.

CLIMATE CHANGE

The science around climate change is not controversial at the most fundamental level. Science has now established to a degree of certainty the fact that human-driven climate change is occurring rapidly, with consequences that will radically alter the life systems of our planet. The kinds of change required to avert disaster are also not debated at the most fundamental level. We have to stop and reverse the amount of greenhouse gases in our atmosphere, especially carbon dioxide from the burning of all fossil fuels.

A variety of measures and technologies are available to enable this change, but changes will distribute costs differently to different people. Some people accustomed to great privileges will need to reexamine their expectations. Others seeking more developed standards of living will need to lead change by developing alternatives to the current models. ***See also* Volume 1, Chapter 2: Atmosphere.**

For example, one major concern is carbon emissions and their degrading effect on the atmosphere. As a result, public policies and private practices are changing to decrease carbon emissions, also called "decreasing your carbon footprint." One basic model of carbon reduction that can be applied to most organizations requires reduction of electrical energy use because most of it comes from nonrenewable sources. Companies and individuals also purchase "carbon offsets" in which the money is used in a project that offsets the carbon emitted by the activity.

References

Adger, W. Neil et al., eds. 2006. *Fairness in Adaptation to Climate Change.* Cambridge, MA: MIT Press.

DiMento, Joseph F. C., and Pamela Doughman, eds. 2007. *Climate Change: What It Means for Us, Our Children, and Our Grandchildren.* Cambridge, MA: MIT Press.

Emanuel, Kerry. 2007. *What We Know about Climate Change.* Cambridge, MA: MIT Press.

Gore, Albert. 2006. *An Inconvenient Truth: The Planetary Emergency of Global Warming and What We Can Do about It.* Emmaus, PA: Rodale Press.

McIntosh, Roderick J., Joseph A. Tainter, and Susan Keech Mcintosh. 2000. *The Way the Wind Blows: Climate Change, History, and Human Action.* New York: Columbia University Press.

Parry, M. L., O. F. Canziani, et al., eds. 2007. *Climate Change 2007: Impacts, Adaptation and Vulnerability. Contribution of Working Group II to the Fourth Assessment Report of the Intergovernmental Panel on Climate Change.* Cambridge, UK: Cambridge University Press.

Rappaport, Ann, and Sarah Hammond Creighton. 2007. *Degrees That Matter: Climate Change and the University.* Cambridge, MA: MIT Press.

Volk, Tyler. 2008. *CO2 Rising: The World's Greatest Environmental Challenge.* Cambridge, MA: MIT Press.

DEMOCRACY AND SUSTAINABILITY

Can democracies act quickly enough to respond to human-generated threats to our ecological life support? Basic democratic theory requires the participation and consent of the many that are governed in public law and policy. Exclusion and lack of consent make government action illegitimate. For those who see the need for change toward sustainable behavior as urgent and widespread, the need to consult widely and to develop political constituencies and leadership for sustainable change seems too complex, indirect, and compromised to promise hope of effective change. Some have argued that timely change on the scale that we would need to avoid collapsing our webs of life can be achieved only by a different arrangement of power, such as a plutocracy or oligarchy of properly motivated people or organizations.

One possible response to this argument is that the time taken to ensure genuine participation on a local level also ensures compliance and eliminates the need for costly and slow enforcement mechanisms. This is one reason that Agenda 21 recommends that government action toward sustainability be formulated at the lowest effective level of government. Another response lies in the observation that effective action need not mobilize all of the population. Effective changes can be generated by smaller groups that exercise symbolic influence if not actual power. The power of the tipping point can be exercised voluntarily even in the absence of governmentally sanctioned power to coerce. ***See also* Volume 1, Chapter 3: Role of Government in Sustainability.**

References

Baber, Walter F., and Robert V. Bartlett. 2005. *Deliberative Environmental Politics: Democracy and Ecological Rationality.* Cambridge, MA: MIT Press.

Hester, Randolph T. 2006. *Design for Ecological Democracy.* Cambridge, MA: MIT Press.

ECOTOURISM

One area of environmental decision making that creates controversy around sustainability is ecotourism. The main idea of ecotourism is to market the natural resources of a place to tourists. This is a form of economic development that benefits hotels, restaurants, and transportation industries. It can provide a revenue corridor for foreign investment that increases capital flow to national economies.

The International Ecotourism Society directly promotes ecotourism and sustainability. It has more than 90 members and creates partnerships between various stakeholders. It sponsors events, training, and education. One of its primary stated goals is to promote knowledge about environmental conservation. There are other ecotourism groups with similar goals.

The controversies start with issues of the impact ecotourism has on the environment, on indigenous peoples, and on the local population. There are concerns that ecotourism may decrease the biodiversity of sensitive environments. Increased exposure to people may alone affect the way plants and animals interact in their environment. The increase in waste that often accompanies more tourists may change the behavior of some animals that find it easier to live from wastes than to hunt food. The propensity of tourists to take souvenirs may decrease the actual biodiversity of an area. In some cases, the tourists and the guides may not be fully informed of the environment they are touring and act in ways that have unintended consequences on a given ecosystem. This is the case in some sensitive tropical rainforests where not all species are yet known. There is some concern that ecotourism may expedite extinctions by direct impacts and indirectly by introducing invasive species or diseases that eradicate native species.

FIGURE 1.18 • The first ecotourist destination, the Galapagos Islands were studies by Charles Darwin and inspired his theory of evolution. Tortoises rest in their enclosure in the captive breeding center of the Galapagos National Park in Puerto Ayora, Galapagos, Monday, January 5, 2009. The islands' local population is seeking ways to take a larger part of the tourism revenue that goes mostly to foreign companies. AP Photo/Kirsten Johnson.

The relationship between indigenous people and government of the nation may be tenuous. Many governments in areas populated by indigenous people seek economic development through ecotourism. One example of some of the dynamics of ecotourism is in Malawi, Africa. The largest lake there is Lake Malawi. Fishing from these fresh water areas is the main food source for this very poor country. The biodiversity of the fish population is very high. On Lake Malawi is the fishing village of Mdulumanja. In this town is a large tourist hotel. The land, once owned by the country, was sold to a private corporation in 1987. The new owner insisted on tearing the village down to expand the hotel. He then stopped allowing the residents to use his water sources and also stopped buying fish from them. He put up a big fence all around his property. In 1990, after much dispute, he evicted the residents with the help of the national government. More than 70 long-term residents were forced to leave. The hotel owner razed all their homes for tourist development. He expanded his hotel to accommodate the tourists who were mainly white South Africans. The residents lived on the fish from that lake and now had to find other means without access to the lakefront.

Ecotourism exploitation is dynamic. Lake Malawi National Park is a world heritage site and may be under the same threat of ecotourism development as the hotel industry marches along the waterfront. There is no monitoring of the impact of this ecotourist industry on the aquatic biodiversity.

The type of job creation under ecotourism is generally in the service sector. Maids, cooks, maintenance workers, drivers, and guides who serve the tourists follow the initial jobs in construction. These are generally low-wage jobs but in places of poverty are considered to be better than nothing. In some places they may exploit cultural and social oppression. The people trained and hired for these jobs may not be from the area. Tourists may not be aware of local cultural dynamics and can unintentionally further political and social oppression. In nations where the indigenous peoples have separate interests, ecotourism may occur as economic development that disrespects native cultures.

Many of the construction practices around ecotourism may not be sustainable and may have negative environmental impacts on natural systems on which present and future life depends. In places where indigenous peoples have interacted sustainably with their environments in holistic ways for long periods, important environmental knowledge could be lost.

References

Dowies, Mark. 2009. *Conservation Refugees: The Hundred-Year Conflict Between Global Conservation and Native Peoples.* Cambridge, MA: MIT Press.

Lumley, Sarah. 2002. *Sustainability and Degradation in Less Developed Countries: Immolating the Future?* Surrey, UK: Ashgate.

Mowforth, Martin, and Ian Munt. 1998. *Tourism and Sustainability: New Tourism in the Third World.* Andover, UK: Routledge.

ENERGY SOURCES

In terms of sustainability, energy is divided into two categories—renewable energy sources and nonrenewable energy sources. Coal, oil, natural gas, and uranium are nonrenewable energy sources because they will run out and are therefore considered nonsustainable. Also, their by-products are carbon dioxide and radioactive waste, which are not readily reabsorbed back into natural systems without damage to the environment. As our supplies of these nonrenewable energy sources run low and population and energy demand increases, retrieving and transporting these energy sources increase cost, international conflict, and environmental impact. For example, drilling for oil in the Arctic National Wildlife Preserve becomes more controversial as the international price of oil increases.

Many controversies exist around these traditional energy sources. Some claim that new technology will make radioactive waste nonhazardous. Others claim that oil can be retrieved and recycled from current uses. Besides energy sources, nonrenewable energy sources such as oil also form the basis for many products that whole societies are based on. Others claim that carbon dioxide can be captured and sequestered, decreasing its effects on global warming and climate change.

Renewable energy sources are considered sustainable. These include the sun, wind, and tides. These sources do not have waste products that harm the environment. Other energy sources are between renewable and nonrenewable. Alternative fuels such as ethanol from corn, biomass converters, geothermal heat wells, and hydropower from dams are sources that may be more sustainable than traditional nonrenewable energy sources. Some dispute this claim because these sources often require more energy than they produce and can produce waste products or activities that have negative environmental impacts.

Aspects of energy conservation are currently major parts of some sustainability programs, such as green buildings. Decreasing energy use does decrease environmental impacts, but if the energy source is not renewable, then with population growth it may not be sustainable. This is the subject of some controversy. With green buildings and sustainability certification programs, energy conservation can mean using natural lights, water-permeable paved surfaces, tree shading, and provisions for alternative transportation.

Reference

Weiss, Charles, and William B. Bonvillian. 2009. *Structuring an Energy Technology Revolution*. Cambridge, MA: MIT Press.

Unsustainable Energy Sources

Contemporary economies are dependent on petrochemical and fossil fuels: oil and coal. These energy sources have significant costs for the environment including spills, dumping, drilling, and mining. Industries that rely on technologies built around these sources of energy will need to make substantial investments in technology and personnel if they are going to change. This scale of change is hard to imagine in the absence of a widespread disaster or war.

Industries built around these fossil fuels employ vast numbers of people. If these industries change, those workers will need jobs in other areas. The transition from unsustainable energy sources to more sustainable ones may cause serious hardship among people and communities employed or reliant on oil and coal.

Among the many alternative energy sources that could replace oil and coal, some choices have controversial impacts as well, such as nuclear energy. The waste associated with the creation of nuclear fuel poses serious risks and has not been resolved in a way that is ecologically sustainable. Nuclear waste is still not a closed ecological loop with the waste providing the basis for future processes. In addition, radiation poses harmful hazards to workers and to communities reliant on this form of energy. Several new types of nuclear reactors, however, may improve their efficiency and decrease the amount of waste. One new design offers pressurized water as a safety factor to cool the reactor if the plant loses power. Another approach uses fist-size graphite balls filled with uranium dioxide to perform the same safety function. There is another proposal that would greatly decrease waste but is considered only theoretical now. It is called "traveling wave" theory. Basically, this uses and reuses spent nuclear fuels in a wavelike set of chemical reactions.

Sustainable Energy Sources

Many alternatives sources of energy are available. Most are generated by local conditions, rather than importation of a commodity. A few offer sufficient power to be exported to other locations. Their generation and use for energy may raise local issues of concern. They all operate without the creation of waste and pollution, based on current knowledge.

Solar Energy

About 85,000 terawatts of energy come from the sun to the Earth every year. Currently about 12.4 gigawatts are captured. This sunlight can be turned into electricity by solar energy technologies. When sunlight hits a solar panel, the energy in the sunlight frees the electrons in the solar cells. This produces an electric current. In the usual crystalline silicon cell, the electrons are tied tightly to atoms. When sunlight hits them, the electrons move more readily, creating electricity. The current power efficiency of crystalline silicon cells is such that about 20 to 25 percent of the sun's energy is converted to usable electricity. Currently mass produced solar cells have only 15 percent efficiency. The highest efficiency solar cells are on satellites and are almost at the 50 percent range of efficiency. Here on Earth the highest efficiency of converting sunlight into electric power is 31.25 percent, set in 2008 at Sandia National Laboratories and breaking the old record of 29.25 percent set in 1984. There is much less interference from the atmosphere and more direct sunlight at higher levels in the atmosphere.

Access to sunlight is restricted by natural cycles of light, darkness, and local weather conditions. This source of energy is free, although the technology to convert it to human uses on a scale useful to modern appliances and fixtures can be a significant cost to business and individuals. Right now the costs of mass production are prohibitive because the raw materials and manufacturing processes are costly. New development in low energy devices (LEDs) such as lights can be applied with solar energy in poor areas of the world to provide light in otherwise unlit human habitations. There is much research in this area, especially in the use of plastics and nanoparticles.

Urban landscape design, building codes, and environmental land use planning can help facilitate the access and use of solar energy. If buildings and other structures block the sun, then access to direct solar energy is denied. Buildings need to be built to hold solar panels and to gain the most from sun exposure. Environmental land use planning can help maximize access to sunlight by requiring alternative energy use, as occurs in parts of Germany.

In terms of environmental sustainability, solar power holds great potential. It is likely that further technological advances in solar power technology will increase its efficiency and decrease the energy impact on natural systems. Solar power also has the potential to reduce the environmental impacts of the energy distribution system because it can be produced near the point of use.

HYDROELECTRIC ENERGY

When water flows and falls, it creates tremendous energy. The force of these flows and falls has the potential to provide energy for many human uses. Technology to convert this energy to human use can begin by stopping the flow or fall and rechanneling its energy. Humans and other animals have used a variety of ways to do this by building dams from all kinds of materials. Beavers use mud and debris from their ponds. Humans use cement and metal constructions to capture the most powerful rivers and falls.

When waters have been sequestered by dams, the particulate matter that they transport will drop to the floor, forming layers of gathered materials called sediments. These sediments can concentrate particulate matter from runoff including highly toxic materials. When dams are removed, one consequence is to wash this concentrated sediment downstream to human settlements and to the oceans or seas. Because human settlements tend to form around water flows for the purpose of transportation, blocking these flows often has the known consequence of inundating preexisting human settlements.

FIGURE 1.19 • Plans for tapping water from the Zambezi, Africa's fourth largest river, represent salvation for the dry lands of Zimbabwe and perhaps all of southern Africa. Courtesy of John Walker.

Falling and rushing water can generate substantial energy that can be harvested and used by humans. Water must be diverted and constrained through a technological device that uses or stores its energy. Ancient waterwheels as well as contemporary dams rely on the same technological idea of using water to power wheels to power machinery. The diversion of water and its constraint may create significant local consequences, depending on the scale of such interferences. Dams may result in inundation of whole geographical features as well as cities and towns. Loss of these areas and the human cultures they contain in order to generate power is a controversial choice.

Fresh water is an increasingly scarce resource and an irreplaceable natural system in a world with an increasing need for it by an increasingly consumptive human population. Its use as an energy source when combined with gravity may or may not detract from other uses such as irrigation. Local or small power production from falling water is another way to be more self-reliant, and may have less of an impact on the environment than a traditional energy source relying on nonrenewable rouses such as coal.

GEOTHERMAL ENERGY

Below the crust of the Earth, the Earth itself is in considerable dynamic thermal flux. Sometimes this thermal condition erupts at the surface in the form of volcanoes and geysers. Less dramatically, the Earth's hotspots manifest locally as steam and superheated conditions that can also generate energy sources useful for human activities.

From the perspective of environmental impacts, current usage of geothermal energy seems to have minimal environmental impacts. For the most part, people are simply gathering the heat that would leave the Earth anyway. One issue is how local or regional the energy is. Can heat be converted to energy such as electricity and be used over a long distance? Another issue is how large a human impact a geothermal energy source could sustain and how large a population and ecosystem it could support.

WIND ENERGY

Winds are powerfully driven across the land and water by their dynamic interaction. The power of these winds can be destructive and overwhelming in the form of hurricanes and tornadoes. Some places and conditions of wind, however, are highly predictable and offer the possibility of beneficial human use. Ancient uses of wind power include wind turbines.

Winds often travel in predictable directions at known places. This makes them good locations for windmills. One of the locations currently being explored for a turbine-type of wind power is a large bridge in the city of Portland, Oregon. The current bridge is old and needs to be rebuilt. The wind from the Columbia River Gorge is strong and predictable. An emerging concept that may increase the predictability of wind is the use of lasers on the wind turbines. These lasers would be aimed at the prevailing wind direction and reflect off particles in the air. When winds are less predictable, power production can ebb, and sometimes gusts can destroy the wind turbine blades. Currently, windmills are also explored for use at a residential level. They have been used in rural agricultural areas to draw water for wells. Like solar power, windmills need access to wind and require building and zoning code flexibility. Wind energy may also be more efficiently captured with future technology in windmill design.

TIDES ENERGY

The sun and moon exert powerful forces on the Earth in the form of the rising and falling levels of oceans and some larger bodies of water. Over time, and in certain locations, the force of these changing levels can be harnessed to produce energy for human uses.

Tides rise and fall around the globe twice a day. In places where tides are constricted by land masses, they can rise and fall more than 40 feet. This is a large amount of water and energy. There are times when tides are still, however, so tidal power plants are operational only about 10 hours a day. So far, only about 20 tidal energy sites in the world have been identified, but some researchers are confident that there are more.

The concept of tidal power generation works like traditional hydroelectric power generation. Basically, a dam is built across a river estuary. In tidal energy terms, this dam is called a barrage. It is generally much wider than a traditional dam across a river. The barrage has tunnels built

FIGURE 1.20 • Wind turbines spin under windswept, cloudy skies along a ridge line in the Columbia River Gorge at FPL Energy's Stateline Wind Project on the Oregon-Washington border near Touchet, WA, March 6, 2003. Wind energy will play a growing role in meeting the rising power needs of the Northwest, but it is not controllable and it needs total backup by traditional sources such as hydroelectric dams, according to a report released Wednesday, March 21, 2007, by energy specialists. AP Photo.

into it that move turbines that create electricity. The turbines are turned by the tide coming in and then going out. The only tidal power plant in Europe is in France in the Rance estuary. It is the largest in the world and was built in 1966.

There are some environmental concerns with tidal power generation. They are built and operate in sensitive aquatic ecosystems. They decrease the normal rate of flow of the water and could upset delicate ecosystems there. Fish, amphibians, and birds could be influenced by the flow alterations.

Other ideas about the creation of energy from tides include the application of windmill-like structures as underwater turbines to create electrical energy. They would require high underwater current velocities or engage continuous ocean currents. Unlike traditional windmills, these turbine blades could operate in both directions to capture energy from ingoing and outflowing tides. This idea is still in the experimental phase. The University of Oxford is experimenting with a turbine that looks like a push lawn mower. These turbines would be built to a very large scale. Although they may be as much as 10 percent less efficient, the economies of scale achieved in maintenance would be worth it.

Tides continue to attract researchers interested in developed alternative and sustainable energy sources. They are reliable and contain large amounts of energy. Their impact on systems of nature on which future life depends is not known to be threatening. The impact of moving large amounts of electrical energy through coastline estuaries is not known. Tidal energy may not be viable in places where tides are small, such as the equator.

In Salem, Oregon, a nonprofit organization called the Oregon Wave Energy Trust has begun to explore wave energy sources. The Trust includes state agencies, private law firms, utilities, and universities. It has received $4.2 million for two years of research into wave energy off the Oregon coast. The organization estimates that Oregon's coast has enough energy to provide 10 percent of the population with electricity by 2025. The Oregon Wave Energy trust has applied to the Federal Energy Regulatory Commission for six project permits. Only one has been issued a preliminary permit to date.

Biofuels for Energy Cogeneration

As materials from plants and animals decompose, they release gases and heat that can be used as fuel by humans. Capturing and converting these decomposing materials as gas and heat in a form that is useful to humankind are sometimes referred to as cogeneration of energy.

Biofuels generally refer to ethanol, butanol, and biodiesel. They replace petrochemicals as an energy source and come from plants such as corn. There is concern that the growth of plants necessary for the creation of ethanol will have negative environmental impacts. About 3 percent of our current gasoline supply is ethanol. Corn, for example, requires more nitrogen fertilizer than other crops and runoff of this fertilizer can overwhelm lakes and rivers. It is estimated that it takes three gallons of water to produce one gallon of ethanol from corn versus 2.5 gallons of water to produce one gallon of gasoline. If other plants are used to produce ethanol, some claim the water usage to create one gallon of ethanol can decrease to one gallon of water.

The Energy Policy Act of 2005 requires fuel producers to nearly double sales of ethanol-blend fuel from 2006 to 2012. Currently, ethanol is mostly used as an additive to gasoline in low blends up to 10 percent ethanol and 90 percent gasoline; however, these blends may actually increase air pollution. They can increase ground-level ozone pollution by increasing emissions of nitrogen oxides and volatile organic compounds that help create ozone. These low ethanol blend fuels also decrease the emission of carbon monoxide, which slightly contributes to ozone depletion.

The development of biofuels as a sustainable energy source is dependent on the types of plants used to create the fuels and how the fuels are used. There are many types of plants that can be used to create biofuels, such as wheat grasses and kudzu. Plants could be genetically created specifically to make fuel. For sustainability purposes the question becomes one of the environmental impacts of these plants in the creation, growth, harvesting, and transformation into fuel. Do these plants and processes threaten systems of nature on which future life depends? Other questions of sustainability depend on how the fuels are used and how they impact their environments. Do they simply replace petrochemicals and emit the same pollutants that cause global warming and climate change? These are difficult, controversial questions. Many amateur and professional scientists are working on these issues.

BIOMASS AS RENEWABLE ENERGY SOURCE

Concern about climate change and increased interest in sustainability initiated search for renewable energy sources that are alternatives to petrochemical sources. The use of biomass for power, heat, and fuels is receiving the endorsement of nations like Germany for use as an alternative energy source. Biomass can be composed of many items, some grown especially for energy use, such as biomass forest timber, fast-growing tree species, and sometimes specially grown cereal straw. Biomass can also be the traditional waste products of other industries. This especially interests sustainability advocates who believe in natural capitalism because it closes the loop, making a by-product into a usable product. Biomass can include waste from agriculture, logging, untreated waste wood and sawdust, and some waste from food production. Wood is the primary biomass fuel. As a fuel wood is considered carbon neutral because it can only release the carbon it received as it grew. Wood is used for site-specific heat generation as well as for heat and power in medium-size industrial plants in Germany.

Biomass in the form of split logs, wood chips, and wood pellets feed ovens and boilers. These systems use electronically regulated combustion systems. Some of these systems use an auger screw to feed pellets for burning. These systems produce much lower emissions than stoves and fireplaces. In many places, reliance on woodstoves for heat pollutes the air with particulate matter resulting from inefficient burning. Larger biomass heating systems used to supply several homes are usually fed with wood chips from machine chipped wood. In larger facilities, biomass in the form of wood is used to create both heat and to generate electricity. The simultaneous generation of heat and electricity increases the overall efficiency of the biomass consumed. The modern wood pellet boilers clean themselves automatically. Ash disposal and servicing are done about once a year.

Germany passed the Renewable Energy Sources Act and market development of biomass increased. In 2006, approximately 70,000 pellet boilers and ovens were installed in Germany. In that year, 160 biomass plants supplying 960 megawatts were operating. More than 1 million tons of wood pellets were produced in 2007. In larger heat generation systems more than 1,000 biomass heating plants generate heat for residential areas and for public properties. Larger wood fired boiler systems are used by logging company for generating heat for commercial and industrial land uses. The Renewal Energy Act provides incentives and rules for burning biomass in highly efficient, small-scale heating plants that burn pellets, split logs, and wood chips. This helps the private business sector avoid expensive startup costs that can be incurred when switching to alternative energy sources. A major part of this incentive is that there is a guaranteed price for the energy produced by biomass-generated electricity that is fed into the public national grid. This price is guaranteed for 20 years. The national grid is required to give priority to accepting this kind of electricity and to purchase it with the guaranteed price. Prices are not static; they change depending on the technology, system size, and natural resources used. The incentive program gives additional money for combined biomass-generated heat and power generation, and for the use of new innovative energy technologies. The German law is now being analyzed as a potential energy policy for sustainability.

Biomass is considered to be the largest form of renewable energy. It provides about 14 percent of the world's primary energy supplies. In the United States, biomass energy contributes 3 percent to the nation's energy supply. The European Commission has recently begun to research requirements for a sustainability scheme for biomass for energy purposes. Many industries are being consulted in the process. Energy companies, project developers, equipment manufacturers, government services, agricultural and forest industry groups, and environmental nongovernmental organizations have been asked to give their input. Many other countries from outside the European Union are interested in these processes. In 2008, the commission made a number of energy and climate change-related proposals including a draft directive to promote renewable energy and to increase its share to 20 percent by 2020. The commission undertook to report on requirements for a sustainability scheme for energy uses of biomass by December 31, 2010. This process is designed to give the European Commission industry views on a biomass sustainability approach. They want to develop key principles and criteria to ensure that the use of biomass for energy purposes avoids environmental risks.

Reference

Roser, Dominik et al. 2008. *Sustainable Use of Forest Biomass for Energy.* New York: Springer.

New Sources of Energy

It is possible that energy sources not imagined could exist and be the source of sustainability, or at least unlimited energy. The energy needs of the world's cities are expected to increase even more than the increases in population. Professional and amateur scientists are exploring ways to create new energy sources in their laboratories and even in home garages.

ALGAE ENERGY

The development of ethanol from corn started investigations into the use of other sources of biofuels. A home scientist named James Sears began experimenting with ways to mass produce biofuels to make them economical and accessible. By using algae's ability to photosynthesize, he may have succeeded.

Algae require water, sunlight, and carbon dioxide to grow. Some scientists are experimenting with ways to make algae grow faster, or bigger. There are many types of algae, over 100,000 strains. Each strain has a different amount of carbohydrates, proteins, and oil. As algae grow they produce oil that can be easily converted into biodiesel and ethanol by fermentation of the algae's carbohydrates. Algae grow very quickly if the right conditions are met and can be harvested much more frequently than corn. Promoters of algae say it can produce more than 10,000 gallons of oil per day, much more than any current plant sources.

Algae production of biodiesel and ethanol may not necessarily decrease environmental impacts on the air if rates of fuel usage increase. To produce algae requires use of brackish water and some require large amounts of plastic. There is also some concern that if a type of algae is genetically created and escapes into the natural environment, it could cause ecological damage. Research is at an early stage, but many home-based entrepreneurs are forming companies to develop this product, pursue patents, and develop new markets.

NUCLEAR FUSION

Of all the energy sources known today, nuclear fusion is one that would reduce carbon emissions to zero and thereby decrease global warming and the rate of climate change. Today nuclear energy is produced through nuclear fission, a process by which atoms are split. Nuclear fusion occurs when atoms become fused, releasing a huge amount of heat energy. The sun is powered by a sea of nuclear fusion reactions. Nuclear fusion has been replicated here on Earth with the highest temperatures in the known solar system. It cannot be sustained very long, however, because it is so hot. It does create nuclear waste but unlike the nuclear waste from nuclear fission, which can last 20,000 years, nuclear fusion waste has a half-life of between 10 and 12 years.

Currently it takes more energy to create a nuclear fusion reaction than it produces, but international research is underway to develop ways to control the heat and sustain the fusion reaction. This project is called ITER (International Thermonuclear Experimental Reactor). It is a large international project under construction to study the plasmas in conditions that will produce electrical energy. The members of ITER are the European Union, the Peoples Republic of China, the Republic of Korea, the United States, Japan, India, and the Russian Federation. The ITER reactor will be constructed in a town in southern France called Cadarache. The energy goal is to produce 500 megawatts of fusion power for extended periods of time. If current computations are correct, this is 10 times the energy that is necessary to keep the fusion reaction going.

There are grave concerns about nuclear fusion. Some fear that it represents overreliance on a technological fix. Others fear that it could set off unknown catastrophic consequences on Earth. Others think the scale and cost are so great that not enough energy can be produced quickly enough for a burgeoning world population.

ANIMAL AND HUMAN WASTE

Humans, like all other animals, produce waste. Waste has an undeniable effect on ecosystems and, when concentrated and untreated, it may pose a threat to systems of nature on which future life depends. As the human population increases and becomes more urbanized, waste treatment becomes a higher priority. Left untreated and ignored by social stigma, human waste can accumulate in parts of nature. Waste treatment and sanitation are hallmarks of a civilized society. Higher levels of human sanitation have contributed 20 years to the human lifespan. Before then, human waste was a vector to many diseases that killed the vulnerable old, young, and weak. Currently, 40 percent of the world's population does not have access to a suitable place to excrete their waste, and 90 percent of the human waste in the world ends up untreated in lakes, rivers, and oceans.

The potential to create energy from waste has been explored and is applied to cattle. Cows produce methane that can be reused as energy. Methane gas is a by-product of animal and organic waste. It is a large part of natural gas. Natural gas is used to power sewage treatment plants and increase the generation of power in other types of power generation systems. For example, it can be used to create steam to power turbines that generate electricity. To use human and animal waste, known as biosolids, the moisture and carbon dioxide must first be removed.

Energy from animal waste often comes from cows. They were first used in U.S. dairy farms in the 1970s when oil prices increased. When oil prices and federal support for alternative energy sources declined, the effort stopped. Now that oil prices are again on the rise, energy deregulation allows the production of electricity from nonutility sources. State and sometimes federal government grants provide assistance to farmers who are again using animal manure to create energy. Traditionally, animal wastes have been moved to a lagoon or holding pond until spring, and then possibly used as a type of fertilizer on the fields. In this process they emit many greenhouse gases. Most of these farms creating energy use some type of "waste digester" mechanism and process. Many of these are homemade and based on the ingenuity of the farmer. The energy content of animal manure alone is low, so it is occasionally mixed with other higher energy wastes. Water washes the waste down into an influent tank where anaerobic bacteria degrade the biomaterial. Cow manure is about 65 percent methane. The process takes about 2 to 3 weeks after which the result is either stored or used in a natural gas engine. This energy is often used close to the point of use, so that the distributional problems of energy creation are lessened. It is estimated that the 1 billion tons of cow manure produced per year in the United States cold generate 88 billion kilowatt per hours of electricity. This is about 2.4 percent of current U.S. annual consumption. It also decreases greenhouse gases by about 99 million metric tons. The material left over is more easily used as fertilizer. Farms also produce food, and to do so while creating energy and reducing waste impacts on the environment is seen as a hopeful development for many advocates of sustainability.

Human waste for energy creation has been used in China. San Antonio, Texas also plans to use human waste for energy. San Antonio is a rapidly growing city in Texas with a shrinking water aquifer. As water aquifers shrink they tend to decrease water quality. San Antonio is seeking to protect its remaining water quality and to harvest the methane from human waste for energy rather than risk the waste further contaminating the scarce fresh water resources. Currently, San Antonio produces 140,000 tons of human waste a year that will be converted into 1.5 million cubic feet/day of natural gas to run its power plants.

Energy from human and animal methane produced in this manner still faces the problems of environmental impacts of energy distribution to point of use. Other methods of waste to methane production, however, can be smaller scale with "bio mass" home converter systems. These biomass "digesters" convert human and animal waste and organic wastes into methane. They use bacteria to break down the wastes and convert it into methane gas that can be used or stored for energy in the unit.

Social attitudes about privacy, defecation, and other aspects of waste do not prevent the impact of waste on the environment. These systems may have other environmental impacts such as carbon dioxide emissions; however, it is an energy production method that can be both centralized and localized, and reduce the flow of untreated wastes into water.

References

Rogers, Heather. 2006. *Gone Tomorrow: The Hidden Life of Garbage.* New York: New Press.

Rosemarin, Arno et al. 2008. *Pathways for Sustainable Sanitation: Achieving Millennium Development Goals.* London, UK: IWA.

DIRECT HUMAN PRODUCTION OF ENERGY

Although not yet practiced on a large scale, some places are beginning to experiment with the direct production of energy from people power. For decades, many parents fantasized about having their children earn the power necessary to watch television by capturing the energy from their exercise. This idea is moving from fantasy to reality as new entrepreneurs begin production and distribution of exercise machines that do just that. Team Dynamo Bikes creates four connected bicycles that generate power. If they are used consistently, the company estimates they will produce about 1.5 kilowatt hours per day. The energy produced by the bicycle riders charge a group of batteries. Some SportsArt 9500 HR trainers, a kind of elliptical machine, can be modified to generate about 75 to 100 watts per hour, leading to a net zero energy usage.

Exercise clubs and gyms often have high energy needs. They have heating, air conditioning, security, laundry, water heating, and equipment energy costs. They compete with each other for members by the fees they charge their clients. As energy costs increase, club owners are looking for ways to lower costs and increase members and profits through alternative energy sources. Using the energy provided by their members is one way. Like other businesses they are also using solar energy.

PIEZOELECTRIC SOURCES

In 1880, Jacques and Pierre Curie discovered the piezoelectric effect. This effect creates electricity when mechanical stress is applied to crystalline structures like tourmaline, quartz, topaz, ceramics, and cane sugars. The amount of electricity produced was proportional to the mechanical stress. First, ultrasonic transducers and quartz timing pieces were produced using this type of electrical energy generation. Currently, most airbags in cars use piezoelectric energy generation to inflate on impact.

There is a great deal of ongoing research on the use of piezoelectric energy. Medical research is investigating it as a source of energy for medical robots. Australian medical research scientists are developing miniature medical robots that use piezoelectrical energy. French researchers are developing materials that generate electrical power from the impact of rain. The Japanese are making mass transit station floors out of piezoelectric materials to generate power. The force of people, a mechanical stressor, on the crystals imbedded in the station floor creates voltage. In Britain, highways are being imbedded with power-generating piezoelectric crystals. It is estimated that one kilometer of road can generate 400 kilowatts of energy.

In terms of sustainability, piezoelectric energy ties energy production to mechanical use. The research is at such an early stage that it is unclear whether it will affect systems of nature in a way that impairs future generations. It does offer an alternative to the use of nonrenewable energy sources for electrical energy generation.

Energy Distribution

The concept of sustainable energy is more than its source. It also includes how that energy is distributed because that has an impact on systems of nature on which present and future life depends. Centralized sources of energy, such as nuclear fusion plants and nuclear reactors, have to be distributed to points of use. Most energy sources are converted at some point into electrical energy for distribution and use. Currently, electrical grids, substations, and lines carry the electricity to the point of use. Generally, the more centralized a source of energy, the more impact it has in its distribution. Electrical wires may create electrical magnetic waves that could disrupt life. Also, the current method of distributing electrical energy loses electricity because of resistance in the line and the transmission. Sustainability, and energy as now conceptualized, could be dramatically different if the wireless transmission of electrical energy becomes a reality. Smaller, localized sources of energy, such as rooftop windmills and solar panels, may be more sustainable because the power is created near the point of use.

Another aspect in the sustainability of energy other than its source and transmission is the environmental impact of the products used in its transmission and creation. Hydroenergy dependent on large dams may produce electrical power but may destroy whole watersheds and aquatic ecosystems. Solar panels with short lifespans and made with toxic materials may fill up landfills and leach into underground water supplies, eventually affecting the land. Nuclear fission energy sources create radioactive wastes that threaten entire ecosystems for tens of thousands of years. A comparison of energy sources requires full cost accounting of the product, its source, and how it is transmitted to the point of use. For example, where electrical power is sent over a grid of power lines or where wind turbines are used, there is concern about the impact on birds and bats. When the birds are an endangered migratory species, as in whooping cranes, the impact is a consideration. Currently, brightly colored diverters are used in some areas to deter birds from the turbines or power lines. Bats are often an important part of local ecologies and consume large amounts of insects. Consideration of all the environmental impacts of all energy sources remains important to full cost accounting in sustainable development.

ENVIRONMENTAL REPARATIONS

In the United States environmental regulation to protect the environment is fairly recent. Although parks have been formed at national, state, and local levels since the early 1900s, these protected environments were small in comparison to the size of ecosystems. Many of the protections offered by parks are incomplete. They allow many activities that compromise environmental protection. In addition to recreational activities, many parks permit concessions such as hotel and ski developments. Some parks permit natural resource extraction such as mining and logging. Other parks permit hunting and foraging. These activities do not protect the environment. The U.S. Environmental Protection Agency (EPA) was formed in 1970. Most states did not have any environmental protection agencies before then. Oregon had the first Air Agency in 1959, but it had very limited regulatory power. State and federal environmental agencies provide limited and recent protection of some environmental activities. Before then, the environment had virtually no protection in the face of rapidly rising industrialization and population growth and expansion. The impacts on the environment were severe with long-term and global consequences.

Reparations are acts that repair damage done by past wrongs. Reparations are always controversial because they require addressing the expense of past actions with present day actors who may not feel responsible even though they have benefited from the past actions. Past environmental acts, whether intended or not, have created damage to systems of life on which future life depends. Some of the past environmental acts may not have been large enough to create this type of damage at the time but do so now. Modern environmental regulation reaches only present-day actions and may not be enough to protect the environment in a sustainable manner. Reparation for sustainability reaches past acts and attempts to mitigate their influence on current and future ecosystems.

Reparations conceptually are often applied in a political context to undo the harm of past human oppression and inhumanity. The arguments for reparations to African Americans for the damages done by slavery are well developed. They include the long-term pervasive differences in African American health, income, education, justice, and housing that are traceable to the conditions of slavery and the post-slavery era. In addition to these data, there is evidence that the stigma that attached to race because of slavery also injured the land and ecology of African American communities. One group of African Americans most involved with the environment was black farmers.

Before industrialization, farming had the most impact on the environment. Forests were cut down, rivers rerouted, roads built, and wildlife decimated so that farms that grew food, cotton, and tobacco could be built. Slave labor on southern plantations was the basis of early U.S. agricultural economy. Most of these slaves were imported from Africa. When slavery ended after a bitter civil war, former slaves were promised 40 acres and a mule to continue farming. The basis for this promise was General Sherman's Special Field Order Number 15 on January 16, 1865. He allowed the freed slaves to continue farming the land of their former owners. In March 1865, the U.S. Congress allowed General Sherman to rent the slaves the land and give them plow mules. Four million former slaves were freed, but they lacked food, housing, and independent community infrastructure. In June 1865, 40,000 freed slaves were leased 400,000 acres of farmland in the southeastern United States. The rent they paid supported the Freedmen's Bureau, which helped create the necessary independent community infrastructure. In May 1865, after President Lincoln was assassinated, President Andrew Johnson ordered General Sherman to return the leased land to its former confederate owners. Nonetheless, some freed slaves did own farms. Land ownership then was needed for one of the important civil rights in any country—the right to vote. If one did not own land, one could not vote, although there were other substantial obstacles to voting. Farming also meant what it meant to all farmers—self-sufficiency; one did not have to rely on others for food.

Farms are necessary for any society or nation to thrive. President Lincoln established the Department of Agriculture in 1862. The department created more than 2,500 offices called the Farmers Home Administration to help agricultural communities. The Farmers Home Administration in the South worked mainly for large white farmers, many of whom had owned the slave-based plantations. Black farmers did not have access to these necessary services. As a direct result, black farmers lost land. More than 50 percent of African Americans in the 1920s lived on farms as opposed to about 25 percent of white farmers. In 1920, African American farmers were 14 percent of all farmers and farmed about 16 million acres of land. In 2003, African American farmers were less than 1 percent of all farmers and farmed .003 percent of the land. Today there are only about 17,000 black farmers. There have been acrimonious demonstrations in Washington, DC, failed lawsuits, and racially discriminatory commercial lending practices. As small farms became big farms, and as big farms became agribusinesses, black farmers have

Environmental Reparations Districts

In many areas the ecosystem is so damaged that regulation of present environmental impacts will not bring the area back to sustainable levels. This is true in cities and in rural areas. One proposal is to create environmental reparations districts.

Environmental reparations districts are modeled after historic preservation districts. Historic preservation districts are strictly regulated areas with definite physical boundaries. The owners of private property within historic districts must get prior approval from a governing architectural review board before making any change in the property. All properties must comply with the architectural types in that district. In a nation like the United States, where private property is equated with individual liberty, such regulation could be considered a "taking" of private property, which would require the government to pay for the property and would generally be too expensive and unpopular to do. In historic architectural districts, however, the courts have decided that

this is not a taking of private property because it contributes to the "average reciprocity" of property value in the historic district. There are more than 33,000 historic districts in the United States, such as the French Quarter in New Orleans, Louisiana.

Environmental reparations districts would be designated areas with a history of high environmental impacts, such as some urban areas. All development would be subject to review by an environmental preservation review board. This board would be comprised of environmentalists, community members, and others. They would review proposals for their impacts on the environment. Arguments about whether this restriction on private property development is unconstitutional would persist. If, however, the average reciprocity of the district is improved through environmental protection with a goal of sustainability, courts could allow it.

Some states, such as Maryland, are experimenting with similar concepts. Maryland is developing environmental benefits districts that would be able to get state funds for environmental protection.

been excluded without remedy or reparation. Some argue that small family-owned farms can be more sustainable because they are closer to the land. If this is the case, then the loss of black farmers means a loss of potential sustainability.

Reparations for impacted land and communities are based on a physical and practical rationale that parallels sustainability. Environmental reparations are based on recognition that heavily burdened systems require repair to be able to sustain all forms of life, regardless of the reasons why these places were impacted. Environmental reparations to unsustainable farmlands and watersheds will benefit communities of color and poor communities, as well as revitalizing the systems on which all living things in those bioregions depend. *See also* **Volume 1, Chapter 2: Human Values and Goals.**

References

Gilbert, Charlene, and Eli Quinn. 2000. *Homecoming: The story of African American Farmers.* Boston: Beacon Press.

Williams, Juan. 2006. *Black Farmers in America.* Lexington: University of Kentucky Press.

EX SITU CONSERVATION: MUSEUMS, ZOOS, AND CONSERVATORIES

Conservation generally means that the environment is preserved in its most pristine and natural condition. Ex situ conservation takes important representative parts of that environment and preserves them. Many plants are conserved this way. Botanical gardens are often local examples of ex situ conservation.

The purposes of ex situ conservation vary. They can be to rescue endangered species, decrease wild collecting that could threaten biodiversity, create life forms for research, save species that cannot be maintained under current circumstances, produce educational materials, and produce materials for habitat restoration.

Ex situ collections can be living organisms, seed banks, pollen, vegetative propagules, and tissue or cell cultures. Strict methods are usually developed to manage these collections. Sometimes these management systems work with conservation management systems that work together to protect and preserve ecosystem integrity.

Ex situ conservation was an important part of survival in human history because protection of the seed banks was necessary to ensure a crop for the next year. In countries at war, protection of the seed banks was difficult because starving people want to eat the seeds to survive. Recovery from war often meant being able to grow new crops, which is difficult without some saved seeds.

Many seeds are developed from that location and are naturally resistant to pests, weeds, and weather conditions. Genetically modified seeds may be more prone to local pests and conditions and require more pesticides that could negatively affect other parts of the ecosystem.

In terms of sustainability, ex situ conservation is very important. It preserves valuable components of the ecosystem and may help mitigate potentially irreparable damages to natural systems on which future life depends.

References

Botanic Gardens Conservation International. www.bgci.org/plant_search.php.

Guerrant Edward O. Jr., Kayri Havens, and Mike Maunde, eds. 2004. *Ex situ Plant Conservation Supporting Species Survival in the Wild*. Covelo, CA: Island Press.

GREEN CONSUMING

Developed countries consume far more of the world's resources per capita than do lesser-developed nations. This leads some to argue that overconsumption (together with waste and pollution) is a greater threat to sustainability than population growth. Within developed nations, as more individuals and companies become aware of the effect that their use of energy and products has on the ecology of the planet, they have responded by trying to make and use products that are less wasteful or polluting. As demand for these kinds of products increases, more alternatives become available, especially when large companies and organizations like governments require their contractors and suppliers to provide these kinds of products.

There are many questions about this kind of consuming, including whether the advertised goods are genuinely less wasteful or polluting when the entire cost of production and product life cycle is considered. In addition, there is suspicion that if the problem is overconsumption as a cultural and moral value, encouraging more consumption is not going to lower impacts and environmental footprints in consumerist cultures. *See also* **Volume 1, Chapter 2: Human Values and Goals.**

References

Dauvergne, Peter. 2008. *The Shadows of Consumption: Consequences for the Global Environment*. Cambridge, MA: MIT Press.

Princen, Thomas et al., eds. 2002. *Confronting Consumption*. Cambridge, MA: MIT Press.

GROWTH AND DEVELOPMENT

Is human growth and development sustainable? Growth in human population and human consumption of resources requires future growth and further development. In some countries, the need for development is based on the need to provide food, shelter, and sanitation for their people. In other countries, the force pushing development is largely based on desires driven by consumer trends and social or cultural practices. The challenge of sustainable human development is whether it is possible to develop and grow within the limits of our ecosystems. Is it possible to meet current needs by using ecosystem services without destroying the potential of these systems to supply the needs of future generations?

If development in underdeveloped countries takes place in the same way as development during the 19th and 20th centuries, the consequences for the environmental webs of life are forecast to be devastating. The additional contributions to the problems of pollution and waste from these additional populations will collapse certain ecosystems, destroying plant, animal, and human life with them. This forecast questions the type and trajectory of development for the 21st century.

The financial investment necessary to change the type and trajectory of development in the next century in both the developed and developing worlds is substantial. These capital investments in technology and infrastructure have been made before as intentional public policies of government. During the ages of imperialism, monarchs and popes gave immunity in exchange for a percentage of the spoils of conquests. In the current age of corporate globalization, governments have protected the investments of corporate shareholders by limiting their liability for corporate debts, a concept called limited liability.

Immunity sometimes breeds impunity among the privileged. Immunity from the consequences of one's actions can create behaviors that would be criminal if done by other than privileged people. The environmental movement is often accused of elitism because many of its members are from privileged areas of society. The costs of some aspects of environmentalism, such as loss of jobs in logging, the price of organic foods, and the cost of sustainability, are ignored by these environmentalists seeking to protect the environment. The controversy is further escalated by the continued failure of some environmental groups to consider the distributional impacts of environmental decisions. Many environmentalists may consider themselves "color blind" and disallow discourse about the issue as detracting from the main objective of environmental protection. To the extent that sustainability embraces an engaged environmental activism, it, too, confronts these issues of elitism and social class exclusion. Some fear that sustainability is now the "safe" environmentalism because it frames the issue of long-term ecosystem survival without dealing with issues of poverty, race, class, or gender. The concern is that sustainable development is expected to be implemented in a top-down, hierarchical manner, with dominate social elites dictating to others what they must do. Without an understanding of the structural forces in a given society that prevents people from access to decision making, there is also concern that the information necessary for environmentally sustainable development will not develop. Sustainable development is considered to be a place-based type of policy. There may be no one universal policy of sustainable development that works for all ecosystems. Sustainable development policies may need to be as diverse as cultures are.

As sustainable development moves into community-wide implementation in democratic and inclusionary settings, this elitism and exclusion become more apparent. The development of community indicators of sustainable development is now facing the challenge of including equity as one of the indicators.

The problem is that the costs and benefits of moving toward sustainable development will not be carried equally by everyone. Many social problems of an environmental nature will be exposed and need redressing for the benefit of the ecosystem. A large part of these problems occur in politically marginalized communities that currently lack the economic means to address them. Higher environmental standards impose costs both in terms of goods and services, and in terms of labor market demand. Who will bear these costs? It may be that those who can most afford to bear the additional costs of sustainability will be the ones least likely to see any direct environmental benefit. Will the application of the precautionary principle to a manufacturing facility in a low-income, high-unemployment community keep them unemployed? Many politically marginalized communities have much to contribute to sustainable development because they are in the places that need the most environmental repair.

Compounding the problem of environmental elitism in sustainable development is that these decisions will need to be made in the face of scientific uncertainty. Although it is known that ecosystems span communities, the extent that they do so is not always known. If sustainable decisions mean that people with resources and privilege will need to pay for direct environmental benefits in other neighborhoods and without scientific proof, then many controversies will develop.

It may be that some ecosystems are more easily converted to sustainability than others. Some difficult choices will develop. Should the communities that are made sustainable for the least cost be addressed first? Should some communities be forgotten if their ecosystems are too deteriorated? Should communities be free to choose to be unsustainable? Strong values will clash as sustainable development policies unfold in the context of scientific uncertainty, global climate changes, and post-normal science applications. *See also* **Volume 1, Chapter 2: Human Values and Goals.**

Reference

Largent, Mark A., George N. Vlahakis, Isabel Maria Malaquis, Nathan M. Brooks, Francois Regourd, Feza Gunergun, and David Wright, eds. 2006. *Imperialism and Science: Social Impact and Interaction*. Santa Barbara, CA: ABC–CLIO.

INFORMATION

Information is a valuable asset in the pursuit of sustainability. Numerous constraints and controversies surround environmental information.

Information about what is in our air, water, land, food, and communities and "right to know" versus confidential business information or national security is one set of countervailing concerns.

Information is essential in making sustainable choices. Government has the power to assemble information on a wide array of topics in order to perform its duties. Information about the environment is assembled in many offices of government. Sometimes that information is publicly available. Timely public access to environmental information is essential knowledge for residents. Limits are regularly placed on the ability of the public to access information gathered by the government. One significant constraint on public access to governmental information gathered about the environment is the ability of businesses that discharge waste and pollution to shield their activities from public scrutiny to the extent that it involves confidential business information, which is defined as information that the business asserts must remain secret in order to protect its trade secrets. Another significant constraint on public access to governmental information is national security. This involves information that the government asserts must remain secret to protect the nation's own health, safety, and welfare. These assertions of a right to secrets can prevent the true and full accounting for information about what is in our land, air, and water. This is only one set of information constraints.

Information and Knowledge: Essential for Sustainability

The current state of knowledge about the environment is not adequate to know enough information for purposes of sustainability. As more information is learned, our knowledge of past and present environmental interactions and impacts increases. With this increase in knowledge comes increasing population growth and greater environmental impacts. A big question is whether knowledge growth can occur fast enough to mitigate the environmental impacts.

As the number and depth of sustainability stakeholders increase, the information available also can increase. One key concern is how "transparent" this environmental information is. Is the information actually available and understandable to all stakeholders? The answer often depends on whether stakeholders have a right to the information. If they do not have this right, they may not have access to it and it is not transparent. Adequate, accurate, and accessible environmental information is also necessary for meaningful participation.

Another constraint is the lack of important indicators of sustainable policy. Development of sustainability measures and indicators is one of the first steps in most public policy in sustainable development. Adequate, accurate, and accessible environmental information is a prerequisite for the development of sustainability indicators.

References

Felleman, John. 1997. *Deep Information: The Role of Information Policy in Environmental Sustainability.* Westport, CT: Greenwood Press.

Hilty, Lorenz M. 2008. *Information Technology and Sustainability: Essays on the Relationship Between Information Technology and Sustainability.* Gallen, Switzerland: Auflage.

Mitchell, Ronald B., William C. Clark, David W. Cash, and Nancy M. Dickson. 2006. *Global Environmental Assessments: Information and Influence.* Cambridge, MA: MIT Press.

Paarlberg, Robert. 2009. *Starved for Science: How Biotechnology Is Being Kept out of Africa.* Cambridge, MA: Center for the Study of World Religions, Harvard University Press.

International Perspectives on Information for Sustainability

The main international approaches to developing information necessary for sustainability comes from Principle 10 of the Rio Declaration and Agenda 21, section III:

> At the national level, each individual shall appropriate access to information concerning the environment that is held by public authorities, including information on hazardous materials and activities in their communities, and the opportunity to participate in the decision making process. States shall facilitate and encourage public awareness and participation by making information widely available.

Agenda 21 was adopted at the Earth Summit to implement the Rio Declaration. Chapter 40 of Agenda 21, section III focuses on information for decision making. It states:

in sustainable development, everyone is a user and provider of information.

Many international standards on access to environmental information adopt U.S. standards found in the National Environmental Policy Act, the federal Administrative Procedure Act, and the Freedom of Information Act. All these laws have limitations in terms of real access to timely and accurate data. They are ultimately enforced in the courts, and access to the courts can be expensive and slow. Most of the environmental information developed by these agencies is for experts only, and hard for the new stakeholders of sustainability, like the community, to understand. International standards also incorporate environmental information access standards of the Aarhus Convention in Europe. These international standards have little effective force in law.

Enormous advances in communication technology such as the Internet, however, now allow environmental information to be sent to a large number of previously excluded stakeholders. The technological ability to view any and all environments in real time may move nations to rapid development of sustainability indicators. Industries that seek out poor nations to pollute may not be able to hide as readily. Nations that exploit indigenous peoples for natural resource extraction will not be able to hide these acts of environmental degradation.

Reference

Josephson, Paul. 2004. *Resources under Regimes: Technology, Environment, and the State.* Cambridge, MA: Harvard University Press.

The Future Needs of Sustainability for Environmental Information

The development of sustainability indicators need environmental indicators on ecosystems and their relationship to global systems of nature on which present and future life depends. Ecosystem dimensions such as dynamic chemical and physical conditions, inventories of the plants and animals, and the actual impacts of all human use need be to accurate and accessible for the development of useful sustainable indicators. In-depth analysis of ecosystems is necessary. Ecosystem dimensions of food, fiber, water quantity and quality, carbon footprints and storage, soil conditions and shoreline changes, and biomass production and rates of use are necessary.

Environmental information should be available freely, quickly, and accurately. It needs to be monitored all the time, in real time. The concept of environmental information needs to be expanded to include the public health of all communities. Information about the state of a sustainable environment ultimately needs to be considered in all facets of life, in business, at home, and in other areas. Citizens need to have the capacity to develop and understand environmental information, and this is the role of education and civic engagement.

These changes will be expensive, and it is unsettled which stakeholder bears the cost. These changes may challenge current political and economic systems. Nonetheless, rapid advances in technology and environmental concern are pushing the development of accurate and accessible indicators of sustainability. ***See also*** **Volume 1, Chapter 5: Information.**

MARKET-BASED SOLUTIONS

Market-based strategies to achieve sustainability use conventional trading incentives to achieve goals related to sustainability. Buying and selling objects of trade are at the core of market-based systems of current emissions trading programs. Companies are assigned the right to emit a certain amount of pollution by their permits. If these rights were made transferable to others, then buying and selling these rights to pollute could create a market in them. Such rights might be an incentive to companies to reduce pollution below permitted levels.

The net effect of such a market exchange would not necessarily reduce overall inventories of pollution unless coupled with a plan to reduce permitted limits. These are sometimes called "caps." In addition, the effect of transferability may be to concentrate more pollution into already burdened communities. The challenge that such solutions must face is how to incorporate environmental and social consequences that are not quantifiable in current marketplace terms.

The use of market-based solutions for sustainability relies heavily on principles of capitalism. These are discussed elsewhere in this section. Market-based solutions require something to sell to willing, able, and knowledgeable buyers. It could be that a sustainable lifestyle itself is a marketable commodity. Communities with sustainable schools, jobs, and homes may be cheaper, cleaner, and sustainably developed and create demand for such a lifestyle. Social concerns about the environmental impact of products such as cosmetics, clothes, building materials, and food already create a market demand for them if they are advertised as sustainable.

One example of how current market forces are applied to natural resource extraction is how Yvon Chouinard invests in logging. He was the founder of Patagonia, a company that prides itself on its environmental ethic and search for sustainability. Chouinard invested in a unique organization called Ecotrust Forests LLC. This fund is run by the conservation group Ecotrust. Its approach is to buy 1,000- to 5,000-acre land parcels with 20-year-old forests. It then finds local loggers to harvest in a sustainable manner. This means taking out a few trees every year as they reach the appropriate size. It means harvesting them in ways that do not degrade the environment by respecting the canopy, the riparian areas, and the habitat of the native species. This system translates into harvesting no more than 25 to 35 percent of the trees each year and replanting enough trees to replace those that were harvested. Ecotrust hires wildlife biologists who examine the integrity of the ecosystem. Ecotrust retains ownership of the land in contrast to traditional timber harvesting methods. Traditionally, large timber investment groups buy very large land parcels with 38- to 40-year-old trees. They prefer to clear-cut them. This destroys all canopies and can cause erosion and siltification. They then sell the land for whatever they can get for it. Ecotrust owned about 13,000 acres in 2008 and hopes to own about 250,000 acres by 2014. From 2006 to 2008, Ecotrust Forests LCC has made pretax returns of about 9 percent.

Another example of using market forces to achieve sustainability is winemaking in the Willamette Valley, Oregon. Salem, Oregon is the location of Low Input Viticulture and Enology (LIVE), which evaluates and certifies vineyards based on ecologically proven viticulture methods. About 33 percent of the approximately 250 wineries in the Willamette Valley are trying to incorporate sustainability. Some have used biodynamic farming approaches since the early 1990s. Others have built their physical infrastructure from recycled building materials. Still others try to use renewable energy sources such as solar and wind. The environment of the Willamette Valley does not have many diseases or insect problems so that the usual problem of pesticides did not occur. As approaches to sustainability developed in the general viticulture industry over time, they were easily applied to the Willamette Valley.

By treating a natural resource like a renewable resource, it is possible to use market forces to achieve sustainability. Sometimes it requires a radical change in approaches, as in logging. Other times it can mean working closely with systems of nature that are present when the industry begins, as in winemaking. *See also* **Volume 2, Chapter 2: Market-Based Strategies.**

References

Aksoy, Ataman M., and John C. Beghin. 2005. *Global Agricultural Trade and Developing Countries.* Washington, DC: The World Bank.

Bacon, Christopher M. et al. eds. 2008. *Confronting the Coffee Crisis: Fair Trade, Sustainable Livelihoods and Ecosystems in Mexico and Central America.* Cambridge, MA: MIT Press.

Hart, Stuart L. 2005. *Capitalism at the Crossroads: The Unlimited Business Opportunities in Solving the World's Most Difficult Problems.* Philadelphia: Wharton School Publishing.

Heal, Geoffrey. 2008. *When Principles Pay: Corporate Social Responsibility and the Bottom Line.* New York: Columbia University Business School Publishing.

Stern, Alissa J. 2000. *The Process of Business/Environmental Collaborations: Partnering for Sustainability.* Westport, CT: Greenwood Publishing.

Webster, D. G. 2008. *Adaptive Governance: The Dynamics of Atlantic Fisheries Management.* Cambridge, MA: MIT Press.

Yanful, Earnest K. 2009. *Appropriate Technologies for Environmental Protection in the Developing World.* Dordrecht, Netherlands: Springer Press.

POPULATION AND CONSUMPTION

There is great debate about what is causing humankind to overuse our ecosystems. Some believe the root cause is overpopulation. Others believe that overconsumption of resources in the developed world contributes far more to the overuse of our resources and environmental degradation than population. They point to the ecological footprint of developed nations in comparison to developing countries.

Population growth itself is related to human health. As developing countries are able to provide effective public health measures and good nutrition, people survive to maturity in greater numbers, which enables greater reproduction. Population also responds to opportunities for women to obtain an education and opportunities to work outside the home. In developed nations, where such opportunities are more available to women, population rates fall over time. In some developed nations, population rates are below replacement levels that may lead to the need for immigration of young people to maintain adequate work forces and the social benefits that are partly financed by their contributions. *See also* **Volume 1, Chapter 2: Ecosystems.**

PRIVATE PROPERTY: CAN IT CONTINUE SUSTAINABLY?

Governments recognize and protect the idea of private, individual ownership of property to varying extents. To the extent that government or culture recognizes individual rights in land and other property, ownership gives individual owners the power to determine what choices to make regarding that property to some extent. In some places, this power is almost absolute, subordinating every other concern. The more absolute these rights of private property are thought to be, the more an owner's choices may be exercised without knowledge of or regard for the consequences of that choice for the ecology of the place or others within that community. To the extent that government ensures such absolute expectations of rights to property, sustainable choices for environment and for community may be compromised or wholly frustrated. When other interests contradict the interest of a private property owner, controversy is certain to occur.

Sustainability may contradict absolutist views of the rights of private property owners by asserting that interests in the environment and the community's interest in equity deserve equal consideration in decision making. This tension is evident in many controversies regarding waste, pollution, and inequities resulting from human uses of resources like land and water.

Owners of private property often do not wish to be held accountable for the consequences of their choices about waste or pollution that they generate, and that can be (or historically have been) passed along to others, including neighbors, the environment, or distant places and strangers. Economists call the passing of these costs along to someone else "externalities."

Advocates of sustainability insist that those whose actions create waste and pollution should be responsible for the externalities they cause. They advocate mechanisms forcing the generator of waste or pollution to pay the full and true costs of these acts to others and to communities. Mechanisms forcing producers to pay these costs include taxes or fees. To the extent that a producer pays for these "downstream" consequences, they are said to be internalized. In addition, to the extent that producers wish to avoid these costs, they will have an incentive to take measures resulting in more sustainable behaviors, less waste, and less pollution. *See also* **Volume 2, Chapter 4: Real Estate.**

References

Freyfogle, Eric T. 2003. *The Land We Share: Private Property and the Common Good.* Washington, DC: Island Press/Shearwater Books.

Fuchs, D. A. 2003. *An Institutional Basis for Environmental Stewardship: The Structure and Quality of Property Rights.* New York: Springer.

Geisler, Charles, and Gail Daneker, eds. 1997. *Property and Values: Alternatives to Public and Private Ownership.* Washington, DC: Island Press.

RISK ASSESSMENT

Risk assessment is a form of analysis of the probability and magnitude of harm from various events and activities. It is widely used to make decisions. Insurance relies on this computation of risk of harm in making decisions about whether to insure, and if so, how much to charge for insurance. Risk assessment is also used in decisions about development projects, and it is used by governments in budgets and planning activities. Related to the science of risk assessment, risk management determines how to plan for and communicate about risks. Risk perception is a science devoted to examining the qualitative aspects of risk, not simply its quantitative aspects.

Risk assessment chooses what factors and events to consider. In the process of choosing, the assessment may be deliberately constructed to exclude certain factors. Often these choices reflect the desire to simplify the task, but the cost of oversimplification is the heightened risk of coming to erroneous decisions about management and communication.

Risk assessment is a complex and controversial method of decision making. Basic risk assessment follows a single chemical in a single pathway through air, water, or land. The choice of which chemical to monitor often depends on whether it is considered harmful to humans, and this is called a human health risk assessment. The choice of which chemicals to monitor is controversial for several reasons. One is that humans react to chemicals differently; the amount or dose of the chemical may determine if it is "harmful or "adverse." The dose response rate among the public in the United States to aspirin could be up to 1,000 times different. Another reason it is controversial is disagreement about the choice of which chemicals need monitoring. Pesticide manufacturers may claim it is safe, while consumer groups may claim it is not safe. The application of the precautionary principle used in sustainability to a single pathway human health risk assessment would ensure that the chemical does not harm natural systems on which future life depends. That would include present-day humans.

Aggregate risk assessment follows many chemicals through a single pathway through air, water, or land and then adds together the risk. This is also known as the additive principle of risk assessment. It has the same dose response and chemical selection issues as human health risk assessments. Many chemicals interact with each other. Aggregate risk assessment can also mean multiple pathways, that is, it follows one or more chemicals through both air and water and sometimes land. Last, aggregate risk assessment can follow multiple chemicals through multiple pathways, but it always uses the additive principle of risk. Because of the potentially greater threat posed to natural systems on which future life depends, the application of the precautionary principle of sustainability is arguably greater.

Aggregate risk assessment is not cumulative risk assessment. Cumulative risk assessment considers all past, present, and future chemicals; their multiple pathways; and the endpoint of the chemicals at a given point in time. Many argue that only cumulative risk assessment is viable for sustainability. Others argue that it is an impossible task given current states of environmental evaluation and knowledge. Cumulative emissions measure all the chemicals that come from a given place or industrial process. Cumulative impacts measure the effects of all those emissions over time on a given endpoint, such as a human or a food product.

In 1998, the U.S. EPA finished its Cumulative Exposure Project based on the additive principle. In this project they modeled the concentrations of the hazardous air pollutants in 1990 all across the United States. They then combined these measurements with unit risk estimates to estimate the potential increase in cancer risk from multiple hazardous air pollutants. The cancer risks from each hazardous air pollutant were added across pollutants in each census tract to estimate a total cancer risk in each census tract. When uncertainties like synergistic interactions, underestimation of ambient concentrations of hazardous air pollutants, different human vulnerabilities, and changes to potency estimates were considered, cancer risk may have been underestimated by 15 percent.

Ecological risk assessment applies principles of assessing risk using the ecology as a chemical endpoint, as opposed to human health risk assessment that uses humans as the chemical endpoint. Ecological risk assessment is currently mandated in the cleanup of U.S. Superfund sites on the National Priorities List. (NPL home page:www.epa.gov/super fund/sites/npl/index.htm.

NPL locator: www.epa.gov/superfund/sites/npl/npl.htm. Ecological risk assessment follows one or more chemicals through the entire ecology of an area, the land, air, water, flora, fauna, and all life forms. It often requires an endpoint in time because healthy ecosystems are dynamic. It also has controversies of dose response and choice of which chemicals to monitor. Also, the current state of knowledge about ecosystems is incomplete.

All the preceding risk assessment approaches do not usually include actual chemical actions. Most assume that the risk is simply the risk of one chemical added to the risk of another chemical. Most chemicals, however, have some effect on other chemicals, especially in cumulative and ecological assessment risk assessments. Chemicals can synergize, or increase the risk, of another chemical. Chemicals can also antagonize, or reduce the risk of another chemical. Chemicals can interact with their endpoint, whether it is human or ecosystem, to increase or decrease risks. Some postulate this chemical interaction with endpoints may be one reason why there is such a wide variance with the dose response rates.

Alberta Desert Sand Cumulative Risk-Assessment Project in the Oilfields

In Canada, the concept of private property is more limited than in the United States, and provinces own the mineral, gas, and oil rights to their lands. In the north of the province of Alberta, there is oil located in a vast area of so-called tar sand about the size of Florida. This is also an area with biological diversity and indigenous peoples dependent on the environment for sustenance. The tar sands are located in one of the largest boreal forests in the world, valuable among other reasons for its role in carbon sequestration and as bird habitat.

To get to the oil without damaging the environment, some type of risk assessment was necessary. Canada decided to allow oil extraction from the tar sands if the oil companies managed all the risks including those that accumulated.

The province of Alberta, however, has allowed extraction on more than 65,000 kilometers of land without environmental assessment, which was not part of the earlier agreements. Many also consider the tar sands of Alberta to be a large source of greenhouse gases in Canada, emitting an estimated 40 tons of greenhouse gases annually, projected to increase to 141 tons by 2020. The type of oil production in these tar fields used large amounts of water, and many parts of this area are considered a desert. Furthermore, many believe that the tailings ponds, where waste from oil extraction is processed, are too close to the Athabasca River. They cover 130 square kilometers and may be decreasing biodiversity and causing cancer clusters among indigenous people.

Oil production in the desert sands of the boreal forests of northern Alberta presented a unique opportunity to study the management of cumulative impacts on an ecosystem. It demonstrated that the many decisions of a fragmented government approach in a context of an increasing climate change underestimate environmental and human impacts. This is an important lesson for other countries that have less governmental regulation, less accountability, and valuable natural resources.

Chemicals can also be simply inert with one another, or may serve as a catalyst. A catalyst chemical may cause a reaction between chemicals but not be part of the final chemical product.

Risk assessment is a primary tool used to make current environmental decisions, but it is still unrefined, expensive, and controversial.

Public Perception of Risk

To manage risk effectively, risk managers may rely on their power to enforce their decisions involuntarily or voluntarily. Risk perception reveals how people feel about different types of risk. Sometimes these perceptions can seem surprising or even counterintuitive. Perceptions are real and help to determine the nature and quality of behavior when faced with certain types of risk. When certain perceptions are excluded from management decisions, there is heightened risk of noncompliance or resistance to management decisions. Decisions to discount or exclude certain perceptions are often made based on lack of scientific expertise. This may result in popularly discredited management decisions. Sometimes decisions to discount or exclude certain risk perceptions are made based on gender, race, or economic class, with the same results. Voluntary compliance based on commonly held perceptions of risk eliminates much of the need to use involuntary enforcement mechanisms.

RACE AND GENDER DIFFERENCES IN ENVIRONMENTAL RISK PERCEPTION: A BIG DIVIDE

How do perceptions of risk affect your stakeholder's involvement in your chosen environmental decisions? This question is a controversial one, but it will have to be fully addressed as social policies of sustainability become implemented. Even if the perception of environmental risk is not accurate, it can cause physiological impacts on the public. The perception of environmental risk or hazard to oneself or one's loved one creates stress that can lead to heart attacks, hypertension, and strokes.

A stakeholder is defined as a person who has a stake, or an interest, in the outcome of a decision. How a decision is framed can determine who the stakeholders are. In recent history, environmental decisions were framed as those between environmentalists and industry. Currently, environmental decisions are framed to include a public health component, as the results of incomplete environmental decisions have begun to accumulate into more than just degrading ecosystems but to affecting public health. This enlarges the stakeholder pool to include communities, governments, environmentalists, and industry. Under a social policy of sustainability, the stakeholder groups may be even more expansive because of the inclusion of ecosystems and longer term environmental considerations.

The perception of risk may not be the actual risk posed; however, even scientific versions of risk posed may be inaccurate. Even expert risk assessors have been shown to have bias in their conclusions based on how they view the world. Fear of catastrophic risk, like a nuclear accident such as Three Mile Island, may pose little scientific risk but may have a large perception of risk. Dread risks, like cancer, may pose little scientific risk per million people, but the perception of dreaded diseases like cancer can increase the perception of risk. The perception of risk can often depend on whether a person is assigning risk for others or assigning risk for herself or loved ones. When assigning risks to environment or to ecosystems, the perception of risk may be lower without humans than with humans. Under a social policy of sustainability, the propensity to lower risks in nonhuman environments is an issue.

With each new set of stakeholders, and sometimes within traditional stakeholder groups, the perception of risk from active and passive environmental decisions changes. An active environmental decision usually has a proponent with an idea for some type of project or activity that has an environmental impact. A passive environmental decision is one that allows the current way of environmental decision making to continue. Because many past and present environmental decision-making processes are incomplete and have significant environmental consequences, many passive environmental decision-making processes will be revisited under more comprehensive policies of sustainability. The perception of risk in both passive and active environmental decisions will be a significant factor in sustainability policies.

Race, gender, income, and educational level all change the perception of environmental risk one perceives and that one is willing to assign to others and to the environment. These perceptions are the result of many factors, such as a history of environmentally disproportionate risks. (For more discussion of such factors, see volume 3). These perceptions of risk can also be the result of being a member of a privileged social group, whether or not that group acknowledged the privilege. As social policies of sustainability revisit passive environmental decisions and become more inclusive of formerly excluded stakeholder groups, some sustainability decisions can become controversial.

In some research examining 30 categories of risk perception, nonwhites perceived environmental risks as approximately 30 percent higher than whites. Females perceived environmental risks as about 30 percent higher than males. Some have argued this leaves a white male aura of invincibility. White males are a privileged group in U.S. society, and this may be an area of unacknowledged privilege. White males are a dominant group in many areas of traditional environmental decision making, such as the U.S. EPA and as chief executive officers of most industries. Some have argued that it is the lack of scientific training that accounts for the disparity between male and female perception of environmental risks. Research shows, however, that the perception of environmental risk between males and females is still about 30 percent different in samples of males and females with terminal degrees in scientific fields.

If white males made environmental decisions based on low perceptions of environmental risk, then many of these decisions may have to be reopened as policies of sustainability begin implementation. This may include many active and passive environmental decisions of the past. An open question is whether the application of the precautionary principle of sustainability, that a decision is postponed if it could harm natural systems on which future life depends, will reopen past environmental decisions. *See also* **Volume 1, Chapter 5: Risk Analysis.**

References

Beer, Tom, and Alike Ismail-Zadeh. 2003. *Risk Science and Sustainability: Science for the Reduction of Risk and Sustainable Development of Society.* New York: Springer.

Cutter, Susan L. 1993. *Living with Risk: The Geography of Technological Hazards.* London, New York: E. Arnold.

Cutter, Susan L. 2006. *Hazards, Vulnerability and Environmental Justice.* London: Sterling, VA: Earthscan.

Morello-Frosch, Rachel, Manuel Pastor, and James Saad. "EJ and Southern California's Riskscape: The Distribution of Air Toxics Exposures and Health Risks among Diverse Communities." *Urban Affairs Review* 36 (2001):551.

Robson, Mark G., William E. and Toscano. 2007. *Risk Assessment for Environmental Health.* Hoboken, NJ: Jossey-Bass.

Whiteside, Kerry H. 2006. *Precautionary Politics: Principle and Practice in Confronting Environmental Risk.* Cambridge, MA: MIT Press.

Cost-Benefit Analysis Applied to Environmental Sustainability

Considering the relative costs and benefits of a decision is a way of trying to determine the value of that decision. This method of analysis often requires comparing the relative costs and benefits of alternatives and choosing the least costly alternative. In this mode of analysis, costs may be limited to financial costs alone or financial costs in a restricted time frame. Similarly, benefits may be limited to financial benefits alone and restricted to particular individuals. The exclusions of broad definitions of costs and benefits may be made to try to simplify the calculation. It can be especially confounding to perform this kind of calculation when costs or benefits are intangible or incalculable for some other reason.

Another weakness in cost-benefit approaches to environmental decision making is the failure to include the health impacts on the public. These are difficult impacts to quantify, and it is even harder to prove causality. Nonetheless, the effects of an unhealthy environment do exist. In the case of air pollution and environmental regulation of ozone, one researcher noted the following.

> It is clear that the impacts of ozone exposure are grave. The body of evidence that ozone causes chronic, pathologic lung damage is overwhelming. At levels routinely encountered in most American cities, ozone burns through cell walls in lungs and airwalls, tissues redden and swell, cellular fluid seeps into the lungs, and over time their elasticity drops. Macrophage cells rush to the lungs defense, but they too are stunned by the ozone. Susceptibility to bacterial infections increases, possibly because ciliated cells that normally expel foreign particles and organisms have been killed and replaced by thicker, stiffer, non ciliated cells. Scars and lesions form in the airways. At ozone levels that prevail through much of the year in California and other warm weather cities, healthy, nonsmoking young men who exercise can't breathe normally. Breathing is rapid, shallow, and painful.

Curtis Moore, 1988 When this type of analysis is used may drastically undervalue ecosystems and their services precisely because they are so unique and difficult to value. In the absence of any replacement for them, their value is limitless; however, infinite value might paralyze decision making. This kind of problem is endemic to ecological decision making, which affects many of the challenges to sustainability that we face.

Reference

Moore, Curtis. "The Impracticality and Immorality of Cost Benefit Analysis in Setting Health Related Standards." *Tulane Environmental Law Journal* 187, no. 11 (1988): 195–98.

Historical Overview of Cost-Benefit Analysis

The first large-scale U.S. public policy to apply cost-benefit analysis was the Flood Control Act of 1937. Federal funds were distributed to flood risks based on whether the benefits granted by the federal government exceeded their costs. Cost-benefit analysis was applied to global issues in the 1950s when the World Bank used cost-benefit analysis to determine investments in developing countries. "Benefits" are the market value of goods and services received. Ecological benefits were not considered, but short-term benefits in terms of profits and single generation quality of life indicators that were measurable were counted. "Costs" were considered the value of goods and services that were foregone. Costs to future generations of loss of ecological services, such as biodiversity, were not considered.

Cost-benefit analysis as a foundation of environmental public policy is controversial when considered through the lens of sustainability. Most economists consider environmental policy as good public policy when it produces positive net present value. Ecological benefits are often hard to measure, and benefits to future generations seldom have present value. One large problem is how to discount future lives. It is impossible to know how many future lives will exist, of humans or any given species. Therefore, some discounting of future lives is required by cost-benefit analysis of sustainable development. Issues of risky future outcomes are hard to handle under cost-benefit analysis when their probabilities are unknown and the values of future decision makers are unknown. Another problem is that it is hard to value something that is lost forever, such as the extinction of a species. Cost-benefit analysis also presents the problem of how to value something that has intrinsic value aside from human values. Many environmental economists do this based on a "willingness to pay" value. This value is dependent on knowing the tradeoffs between choices now and for sustainable development into the future. So much of the environment is just being discovered that values based on present environmental knowledge may be incomplete.

References

Ackerman, Frank, and Lisa Heinzerling. 2004. *Priceless: On Knowing the Price of Everything and the Value of Nothing.* New York: New Press.

Hunkeler, David et al. 2008. *Environmental Life Cycle Costing.* Boca Raton, FL: CRC Press.

SCIENCE FOR POLICY DECISIONS

Scientists and experts are now being asked to provide guidance about matters on which there is inherent uncertainty. Where ecological controversies exist, some basic facts may be unknown, and the underlying values about actions to be taken are often contested. Science relies on certain methodologies to manage uncertainty in its investigations.

These methodologies do not try to resolve dispute as much as they limit the scope of findings.

Post-Normal Science: Policy in a Context of Uncertainty

Traditional science is often a search for certain truths. The null hypothesis often seeks to confirm or deny a hypothesis within a certain statistical range of certainty. The range of error is generally between 2 and 10 percent. Under traditional scientific approaches, uncertainty means that scientists can neither confirm nor deny causality. If the presumption is in favor of the status quo and burden of proof on those challenging the status quo, then it is difficult to change from traditional policy to sustainable development policies.

Many policymakers face decisions with great environmental uncertainty. Global warming and concomitant climate changes put many policymakers in very difficult positions where they can no longer rely on the old way of deciding issues. There are no risk-free decisions, and more and more decisions evoke controversy and public concern with unintended consequences. Science is no longer reliable as a decision-making back up. It is still an important tool.

Traditional science is called normal science. Post-normal science is the policy consideration and management of complex policy issues heavily dependent on science. Post-normal science seeks to be applicable in areas of great uncertainty and more valuable to policymakers. It engages values and seeks to make them explicit, whereas traditional science seeks to be as value free as possible. Post-normal science engages many different value perspectives in this fashion, which helps make it more applicable in a culturally diverse world with many different values. Unlike traditional scientific explorations for the advancement of knowledge and research for its own value, post-normal science seeks to develop knowledge to solve modern-day problems.

As policymakers face the complex problems of sustainable development, such as measuring cumulative risks on an ecosystem basis, evaluating ecological carrying capacity, and implementing the precautionary principle, they will face a lack of science and increased risks of a wrong decision. The level of uncertainty is very high. A wrong decision could jeopardize the systems of nature on which future life depends and impair sustainable development. Post-normal science seeks to make science engaged and applicable to these very complicated and unavoidable decisions.

SUSTAINABLE DEVELOPMENT

Sustainable development is defined as the ability to meet the needs of present generations without compromising the ability of future generations to meet their needs. Sustainable development assumes continuous economic growth, without irreparably or irreversibly damaging the environment. Human population growth is difficult for this model because it is difficult to place an economic value on the lives that exist in the future. Some environmentalists challenge the assumption of growth at all. The fundamental battleground for this emerging controversy is one of values. The continued prioritization of economic growth over environmental protection combined with population increases may have irreparable impacts on the environment, and therefore the model of sustainable development would require governments to place constraints on developmental that have not been present before.

Countries committed to development in order to meet the basic needs of their contemporary populations face difficult choices for sustainability including the choice of energy resources to use. If these countries choose to develop a fossil fuel-dependent economy, the additional contribution to worldwide ecological crises like climate change will certainly result in more ecological damage. Alternative fuels and technologies s, however, are less readily available and often much more expensive to procure. To develop without reliance on fossil fuels, these countries need access to technologies and funding for investments that offset the costs of those options. *See also* **Volume 1, Chapter 2: Human Values and Goals.**

Reference

Collin, Robert. 2008. *Battleground: Environment.* Westport, CT: Greenwood Press.

TRAGEDY OF THE COMMONS

The tragedy of the commons is a story about what happened to land when it was overused by humans to raise animals. This story says that when people were allowed to graze their sheep and goats on land that belonged to everyone, they put so many animals on the land that the land became overgrazed, and ultimately the animals suffered, too. This story is often used to describe the way in which some complex systems can collapse when left unplanned or managed for the common good.

Some have described this dynamic as a positive feedback loop, where gains around a particular or product tend to create more of that same thing. An example of this is population growth, where there are more people, more people will tend to reproduce and create a growing spiral of population until it becomes unsustainable.

Prisoner's Dilemma: A Sustainable Solution to the Tragedy of the Commons

The prisoner's dilemma is another story told to illustrate patterns of human behavior when we are faced with choices that appear to maximize one person's advantage over another. Four prisoners are given the choice to cooperate with officials holding them in exchange for treatment that is more favorable. If none of them cooperates, they will go free. The best outcome for all of them is to remain silent, but self-interestedness may deceive them into a choice that is not in their best long-term interest. The solution to the prisoner's dilemma is to overcome the impulse to short-term self-interest. This is also one solution to the tragedy of the commons dilemma.

What must be supplied in this story as well as the tragedy of the commons tale is an additional narrative about how the individuals involved can come to see both the folly of their immediate choices, and the desirability of changing them in favor of behavior that benefits all of them whether they are shepherds or prisoners.

Public Lands

Some publicly owned lands are open to managed uses for private benefits. Ranchers may lease public lands for grazing sheep and cattle. Oil and gas corporations lease the right to extract oil from public lands. Mining companies own the rights to mine some public lands for their mineral content. National forests can be logged for private profit. But other types of publicly owned lands are held immune from private profit to a greater extent. National parks are open for tourism, and concessionaires offer tourist-related services like hotels and food services, but they are not open for extractive purposes. These differences arise from very different public policies about conservation and environmentalism, as well as from different geographies and ecosystems.

There is growing tension between private uses that can degrade the environmental systems of an area, and the idea of areas that should not be used or developed at all. Particularly in forests and ocean areas, there is growing concern that private uses of resources in these areas have not been respectful of the interconnected ecosystems and habitat. Private users of public lands now expect that they may exploit the resources of publicly owned lands as a matter of right because they have been allowed to do so in the past. Managers of these lands are under increasing pressure to exclude private uses, even limiting tourism and related services to preserve the integrity of ecosystems for the benefit of nonhuman species.

The Case of U.S. Forests

Forests provide many ecosystem services such as carbon sequestration, water purification, and habitat. Public and private property owners determine the ecosystem fate of most forests in the United States. The economic value of forests lies primarily in the wood for heat, energy, building, and many other uses. Since first contact by European settlers, what became the United States has lost about one-third of its forest cover. As forest acreage rapidly decreases because of development, its ecosystem services become more apparent.

U.S. forests have continued losses since first contact, although 3.6 million acres of forests were gained from 1982 to 1997 because abandoned crop and grazing lands were changed to a forest designation. During roughly the same period, about 10 million acres of forest was destroyed for development. Forest acres were converted to human habitation or workplace at twice the amount of any other land category. According to the U.S. Department of Agriculture, the annual rate of forest acres lost to development from roughly 1993 to 1998 increased by 70 percent over the 1980s. This comes to about 950,000 acres of forestland destroyed for development every year. Forest policymakers are particularly concerned about forest loss in timber-producing areas of the country.

Forest acreage loss is more than acreage of trees. Forests differ greatly. Older forests have greater biological diversity and generally offer more ecosystem services, such as carbon sequestration. Replanted or regenerated new forests do not offer the biological diversity and generally offer fewer ecosystem services such as carbon sequestration. The economic value of timber often pushes the harvest of trees before they are biologically mature. Frequent harvests can also have environmental impacts on the soil through depletion of soil organisms, soil compaction, and habitat destruction. The ecological value of lost forest acreage may not be quickly replaced by new replanted forests or reconverted agricultural land. Developed land is seldom reconverted to forest acreage.

Developed land is land developed for office, residential, commercial, and industrial buildings. Usually it is preceded by a road or a road enlargement. Increasing population and rapid deforestation cause the fragmentation of forestlands. This fragmentation of public and private control of forestlands is overlaid with weak enforcement of environmental and land use laws. Fragmentation can decrease the ecosystem services of forestland. When parts of the forest become separate from other parts, the forest ecosystem begins to break down. Patterns of movement of plants and animals, water and sunlight, and soil and air are all disturbed. Between 1978 and 1994, about 2 million acres of forestland were fragmented into land plots smaller than 100 acres. In 2000, about 32 million acres were held in land plots smaller than 20 acres.

In many ways, forests are a commons. The tragedy of this commons is that it is related to the habitat necessary for all species. Loss of forestland is considered one of the greatest threats to biodiversity. It is estimated that about 40 percent of all U.S. fish species; 35 percent of amphibians and flowering plants; and 15 percent of birds, mammals, and reptiles are vulnerable to extinction.

As forests are lost, the ability of the remaining forests to function becomes a question. Industries that own wood-producing facilities own about 10 percent of the U.S. forests, and other nonindustrial private owners account for 48 percent of private forest ownership. In fact, .3 percent of all owners own almost 40 percent of all forestlands. Most of the owners of forests in the United States seek a short-term cash profit from their trees. They feel they are not compensated for their ecological value, and as private property, owners can harvest their timber. In some cases, forest owners harvest their timber before endangered species, such as the spotted owl, can move and establish a new habitat. Once an endangered species establishes habitat, trees cannot be cut down. Ironically, environmental regulations designed to protect endangered species and lessen environmental impact motivate private property owners seeking cash profits to quickly further degrade the environment. Sustainability challenges the tragedy of the commons and offers policy solutions to it.

One main approach to solving the tragedy of the commons problem in forestry is stewardship forestry. This forestry develops forest management practices that interact with natural forest composition, structure, and process to produce ecosystems services, inclusive of wood. The stewardship forestry approach views trees in their ecological lifecycle.

Trees have produced needed resources in terms of fuel, shelter, and other uses for many years. Logging timber on public and private lands experienced increased efficiency through technological developments in logging equipment and operations. Increased demand for wood products worldwide, increased logging efficiency, and decreasing natural resource base have fueled intense controversies around logging. For example, when endangered species, such as the spotted owl, have a habitat in a timbered area, that area may not be logged. Where the local economy is based on timber extraction and this activity is stopped through the protection of endangered species, controversies flare. Tropical deforestation via logging may be a contributing factor to global warming and climate change because trees there convert carbon dioxide to oxygen. With a worldwide push for sustainable development and in the context of a necessary and declining natural resource, forest stewardship activities increased.

Forest stewardship also affects the secondary markets for wood products. Wood is used for construction but also for many products such as furniture. Market demand for wood in secondary markets from forests certified as sustainable is a large factor in determining market demand for sustainably obtained forest products. Because the secondary market often combines the wood with other products, the environmental market penetration and niche development can be even greater than primary wood markets. In the case of furniture, many paints, glues, and wood treatment chemicals are often used in the production of the item. Many of these paints, glues, and other wood treatment products have volatile organic chemicals in them that can affect both the environment and the health of the user. By combining the market demand for wood from forests certified as sustainable and the market demand for safe, nonhazardous home products, a strong niche in the market is created. If the forest certification and alternatives to toxic chemicals are sustainable, these new market niches may propel forest stewardship to new levels of public policy.

In the United States, the Cooperative Forestry Assistance Act of 1978 authorizes the Forest Stewardship Program. The Forest Stewardship Program provides aid to nonindustrial private owners. Almost half of all forestland in the United States, or about 354 million acres, is owned by nonindustrial private owners. The goal of this program is to create active, sustainable forests with the private owners acting as stewards. It does this by helping owners develop multiuse management plans. Since 1991, the Forest Stewardship Program has produced more than 280,000 forest management plans covering over 32 million acres of forests. Forests under these management plans are expected to provide wildlife habitat, recreational activities, timber harvests, and protect watersheds. The express goal is to maintain that forest for future generations. Private owners must participate in the program for at least 10 years.

Other federal programs also aid in the stewardship of forests. Many states also have forest stewardship programs. Many cities have urban forestry programs with a goal of stewardship. Critics argue, however, that private, voluntary stewardship programs are complicated and do not cover the amount of forest land that watersheds, habitats, and ecosystems need to be sustainable. They maintain that private owners are primarily motivated by profit and not environmental preservation. They point to differences in forest habitat between a forest that is periodically logged and one that is left to ecologically mature. In the case of ecological maturity there is often greater biodiversity, migratory pathways are intact, and watersheds remain stable. In forest habitat that is periodically logged, there is more disturbance in the habitat. This can mean less biological diversity, interrupted migratory pathways, and unstable watersheds. These criticisms are basically the same for stewardship programs on publicly held lands. In the international and indigenous context, the concept of stewardship can have different meanings. In cultures where a person is considered as one and inseparable from the environment, the concept of stewardship can mean to take care of oneself. It can also mean a designation of territory or tribal rights.

The forest as a commons is suffering tragedy. The world is becoming deforested, and it is affecting many ecosystems. Exactly how it is doing this, and to what degree and rate it is doing so is a matter of debate. Private voluntary stewardship programs may not be enough to prevent this problem because they rely on market-based profit motives for present generations.

References

Binley, Dan, and Oleg Menyailo. 2004. *Tree Species Effects on Soils: Implications for Global Change.* New York: Springer.

Moran, Emilio F., and Elinor Ostrom. 2005. *Seeing the Forest and the Trees: Human-Environment Interactions in Forest Ecosystems.* Cambridge, MA: MIT Press.

Robinson, John, and Elizabeth Bennett. 1999. *Hunting for Sustainability in Tropical Forests.* New York: Columbia University Press.

Roser, Dominik et al. 2008. *Sustainable Use of Forest Biomass for Energy.* New York: Springer.

Stein, Bruce A. et al., eds. 2000. *Our Precious Heritage: The Status of Biodiversity in the United States.* New York: Oxford University Press

URBAN SPRAWL

The majority of people on Earth now live in cities and towns. Urbanization allows humans to develop diverse economies that are more resistant to ups and downs. Human settlement has consequences for the ecosystems surrounding and supporting urban areas. Most human settlement is near the coasts and around rivers and lakes. Urbanization often occurs in coastal areas near fragile estuaries and wetlands. These urban areas have grown rapidly in population and geography, often without controls or services. Some of the results of urbanization include the loss of wetlands, farmlands, and forests to paving and construction.

Urban sprawl aggravates several types of environmental problems. Sprawl contributes to air pollution because people often live at a distance from where they work, and public transportation is inadequate as a means of getting to work. This means there are more individual cars and trucks on our roads contributing to poor air quality. Increased air pollution also increases other problems from climate change to human health problems like childhood asthma.

Urban sprawl also aggravates problems of equity and fairness within communities. Sprawl is associated with lack of affordable housing in desirable locations displacing established residents, and the increased development of rural land for suburban housing displacing farmers and farmlands.

Urban areas now occupy about 2 percent of the surface of Earth. The world is rapidly urbanizing, with urban centers become megalopolises. There are now 19 megacities on Earth with more than 10 million residents. By 2015, some predict that 18 of the 27 predicted megacities will be in Asia. The environmental impact, or ecological footprint as it is sometimes called, of urban areas is very large. Cities and their commercial, industrial, and residential uses consume about 75 percent of Earth's natural resources. Cities take natural resources from surrounding ecosystem and biomes. They also produce large environmental impacts that can affect ecosystems outside their region. More than half the world's populations of over 6 billion people live in cities. It is predicted that by 2030, there will be 2 billion new city residents. In 2030, 60 percent of the Earth's population will live in urban areas. Also in 2030, there are predicted to be 500 cities with a population of 1 million of more. This is one reason the international sustainability community focuses on the growth and development of sustainable cities. *See also* **Volume 1, Chapter 2: Land, Soil, and Forests.**

Brownfields: Urban Land Redevelopment

As our cities grow into megolopolian areas with increasing populations, increasing consumption, and increasing environmental degradation of natural systems, urban land redevelopment patterns shift to reclaim industrial land rather than use "greenfield." A strong policy of reusing formerly industrial land can decrease the overall environmental impacts of sprawl by increasing the density of human populations, decreasing trip generation, cleaning up threats to people and ecosystems, and preserving intact ecosystems and biomes. *Brownfields* is a term describing lands contaminated by industrial development. Many brownfields are now located in urban areas where industries chose to locate near crossroads or river transportation. As urban sprawl becomes recognized as a problem, the cleanup and reuse of urban lands become attractive as an alternative to additional sprawl.

Cleanup of brownfields is often expensive. The costs of cleanup vary with the type of toxicity encountered at the site, the geography of the site, and the standard of clean adopted as the goal. Current owners of these sites may have to pay part or all of these costs, even if they are not the ones who made the mess. They have the right to recover their costs from the people or business that did make the mess if they can discover who they are, and if they are still able to pay anything. As a practical matter, this often means that these sites are not cleaned up. To help owners clean up these lands, the U.S. government has agreed to pay some or all of the costs of cleanup through grants and tax credits. The government also assists with brownfields redevelopment in other ways including providing insurance for these large projects.

Abandoned brownfields are often located in blighted communities. The presence of a brownfield is often a significant contribution to the blight. Another challenge involved in the cleanup and reuse of these lands is whether it should be cleaned to a standard safe for residential use, or whether it should restored to a lower standard suitable for other nonresidential use. The surrounding community may not always have a voice in this decision. A private owner of the property might make these decisions based solely on individual best interests. Political accountability for cleanup standards and reuse is greater when government funding is involved.

Sustainable Development and Brownfields

As knowledge about past and present environmental impacts on urban environments increases, so will knowledge about how to develop them sustainably. There is no question that the area of cities that are toxic to people and to life generally must be contained and detoxified. These environmentally toxic areas, or brownfields, may contaminate whole ecosystems threatening other natural systems on which life depends. The extent of these contaminated areas and the rate of ecosystem degradation are currently unknown but thought to be increasing. Ideas such as environmental reparation districts are still theoretical.

A policy of community-based sustainability relates to brownfields in two ways. First, as discussed previously, it cleans up old toxic areas that present irreparable damage to systems of nature on which life depends. It restores the environment to begin to make the ecosystem healthy. The other way brownfields relate to sustainable community development is to prevent the creation of toxic hot spots in the first place. The application of the precautionary principle to traditional land development, the creation of environmental reparation districts, and green building techniques can prevent some of the environmental impacts that led to brownfields. Part of the prevention of Brownfields is the development of community-based sustainability indicators. These include measures of the environmental impacts of the human community on the air, water, and land. Transportation systems, construction techniques, energy conservation, and industrial processes are examined closely for their potential environmental impacts and public health implications.

Urban Ecology

Early uses of the terms *ecology* and *environmentalism* focused on wilderness lands and wild animals to the exclusion of human settlements. Cities and towns were studied by academics and regulated by politics separately and apart from nature and environment. This early disconnection is still evident in the ways in which land use and environmental laws fail to interconnect and in the ways in which environmental agencies often fail to interact with human health agencies. These disconnections are amplified in federal systems by the lack of communication among different levels of government.

As environmentalism and ecology studies have matured, they have had to reengage cities as part of the natural world and landscape. Urban ecologies are complex systems of interactions between humans and their infrastructures and natural systems, plants, and animals. Urban ecologies are also complicated by human social interactions.

To the extent that humans are viewed as a fundamental environmental problem, rather than part of an ecological system, the inclusion of human concerns in the study of environment or ecology is criticized. Some fear that issues of human social interaction like racism and urban politics will detract from the resolution of important environmental issues, or subordinate these issues to other concerns. They reject the interconnectedness of human problems and environmental issues.

WAR

World War II precipitated the exploration of a wide range of chemicals now commonly in use today. Before that war, these chemicals, especially those using the chlorine molecule, were not available. Now they are ubiquitous. Although war tends to drive the chemical industry's research and development agenda, it also imposes catastrophic events on ecosystems, with some unanticipated results.

CHAPTER 5

Future Directions and Emerging Trends

Public policy for sustainable development is fraught with great un-certainty and risk. Post-normal science, ecological risk assessment, application of the precautionary principle, and thorough and accurate environmental monitoring may decrease some of the uncertainty and risks but may uncover others. One public policy approach is to look for trends in decision making around sustainability and for indicators of good sustainable policies. As our approach to sustainability becomes globalized, the approaches to sustainability greatly increase. Many places in the world are trying different methods and approaches to be sustain-able. They are doing so with international aid and assistance and in communication with global networks. Some of these places have similar values, others have very different values, and others have not examined their core values. Most of these places do share one concern, however; they want to develop policy that is sustainable and within the carrying capacity of their ecosystem.

CHANGING OUR THINKING: FROM INEXHAUSTIBLE TO IRREPLACEABLE

One of the most effective ways to achieve change is to change the story that we tell ourselves when we are trying to make sense of complexity. It is human nature to see and think what we already know and what we want. A problem like sustainable development may be more com-plex than anything in our recorded history. The level of uncertainty overwhelms our most trusted leaders, our best scholars, and our bright-est diplomats. It is a dynamic complexity that resists simplification and static description. As cultures try to make sense of the complexity of sustainability, the stories told must also change, requiring changes in traditions and deeply held values.

Religious myths of natural superabundance and infinite providence have a powerful hold on the ability of humans to imagine our relation-ship with the Earth. These creation myths also tend to support the idea of nature's instrumental value to humans, a human-centered approach focused on the utility of nature to humans.

The ability to change the story of our human relationship with our planet, from one of righteously privileged user toward an appreciation of fragile and irreplaceable nature of the dynamics that support life as we know it, is powerful. The power to revise these deeply held intuitive views enlists many other change methods without requiring explicit agreement or instructions. Changing such perspectives requires equally compelling cultural views. Contemporary media from films and television to the Internet have made a powerful counter-message of the fragility of the webs of life on which all living beings depend. Religions have joined in the Earth Charter to try to bridge these conflicting stories and provide the basis for transcending the barriers that confront us in our journey toward sustainability.

References

Clapp, Jennifer, and Peter Dauvergne. 2005. *Paths to a Green World: The Political Economy of the Global Environment*. Cambridge, MA: MIT Press.

The Earth Charter, Earth Charter www.earthcharter.org/.

Edwards, Andres R. 2005. *The Sustainability Revolution: Portrait of a Paradigm Shift*. Gabriola Island, BC: New Society Publishers.

Pirages, Dennis, and Ken Cousins. 2005. *From Resource Scarcity to Ecological Security: Exploring New Limits to Growth*. Cambridge, MA: MIT Press.

Warner, Keith Douglass. 2007. *Agroecology in Action: Extending Alternative Agriculture through Social Networks*. Cambridge, MA: MIT Press.

Scientific Paradigm Shifts

In Ventura, California, an innovative paradigm shift in educational and scientific empowerment is beginning. Coastal Marine Biolabs Integrative Biosciences Program is a nonprofit organization that provides inquiry-based, multidisciplinary learning opportunities for students and educators. They examine many areas of scientific inquiry including bioluminescence, genetic variation in marine populations, bioacoustics, chemical signals in the marine environment, biomaterials, microchemical analysis, phytoplankton biomass and composition, and kelp forest monitoring. Their operating philosophy is one world, one science. In a major shift from current ways of thinking (a paradigm), they operate as if the natural world is all one science.

Barcode of Life Initiative

They are part of an international research effort to barcode the life of Earth by helping their students to do so and thus empowering them with scientific discovery and contributing to scientific knowledge. An extremely important issue in sustainability is knowledge of the species of an organism. Species identification is important to food supply, biodiversity networks, ecosystems, and the economy. Yet standard methods of species identification are difficult, complex, and incomplete. They rely on morphological features to identify biological specimens. The new approach is to use the DNA strands of different species and to translate that into a barcode.

All species have distinct barcode gene sequences. In the last 300 plus years of Western science, taxonomists have collected more than 1.7 million species of plants, animals, and microbes. There are still unknown species. By providing tissue samples, a reference barcode from that gene structure can be produced. The DNA is extracted from the tissue; the barcoded region of the DNA is isolated, replicated by polymerase chain reaction (PCR) amplification. This is a way to make copies of genetic materials. Finally, the DNA is sequenced. It takes a few hours and is very inexpensive. These barcode records contain the DNA sequence, other data on the voucher specimen, and the name of the species. These records are then stored for free access in three locations: GenBank, EMBL, and DDBJ. (See more at the Barcode of Life Initiative, www.dnabarcodes.org.)

DNA barcoding offers many benefits. It helps identify unknown species, risks of extinctions, and paths of probable evolution. It can help identify, monitor, and protect endangered species. The illegal transshipment of exotic species can be enforced inexpensively and quickly at ports by DNA testing. It makes basic research in taxonomy simple, which in turn makes the allocation of resources for basic scientific research easier. DNA barcoding is effective at monitoring population size for species populations because it can be applied at any stage in the lifecycle of a given species. The interaction of species biodiversity with each other in a given ecosystem is often dependent on the stage of lifecycle of a given species. With DNA barcode identification, scientists, government enforcement agencies, environmental advocates, and the public can dramatically increase their knowledge about ecosystem-wide interactions based on biodiversity. Other practical applications include controlling agricultural pests and interrupting disease vector before they spread to large populations. For example, there is a Mosquito Barcoding Initiative that is identifying 85 percent of all known mosquito species. These species are difficult to identify in egg and larval stages. It makes a big difference which species are coming into a region for an effective eradication program that does not harm other species in those ecosystems. By accurate and early species identification, fewer pesticides may be needed.

In 2003, researchers from around the world formed the Barcode of Life Initiative. It is proving to be cost effective as a method of species identification and for government enforcement of species protection. The University of Guelph in Ontario, Canada was the first to pursue barcoding life and now barcodes 100,000 species per year (see www.barcodeoflife.org). The Canadian Barcode of Life network was the first such national network. It engages 50 researchers in 42 organizations. Its stated goal is to gather barcode records for over 10,000 Canadian animal, plant, fungal, and protest species over five years. For more information on this project see Canadian Barcode of Life Network, www.bolnet.ca. The All Birds Barcoding Initiative works with ornithologists globally to barcode 10,000 bird species by 2010. The purpose of these bird barcodes is to monitor birds as an indicator species of environmental quality and to prevent airplane collisions with birds. For more information on this project see All Birds Barcoding Initiative, www.barcodingbirds.org. There is also a global fish barcoding project called the global Fish Barcode of Life Initiative. It plans to gather DNA barcodes from at least five representatives of all 30,000 species of marine and freshwater fish. The goal is to help manage fisheries to levels of sustainable consumption. For more information on this program, see Fish Barcode of Life Initiative, www.fishbol.org. Another barcoding program is ecosystem based. This is called the Biocode Project on the island of Moorea, French Polynesia and is organized between France and the United States at the Gump Marine Station at the University of California at Berkeley. It wants to be the first to get the barcodes of all species in a tropical ecosystem from terrestrial insects and plants to marine invertebrates. For more information, see moorea.berkely.edu/.

Some scientists are pursuing the entire genome legacy of some species. The Ocean Genome Legacy is collecting, describing, and protecting the biological legacy of the oceans. From a sustainability perspective, there is great concern that pollution, overfishing, and resource exploitation is creating irreparable damage to systems of nature on which future life depends. To do this, the Ocean Genome Resource was created as the first public repository for genomic materials, not just DNA. This resource includes DNA, RNA, DNA libraries, amplified DNA, and tissue samples all available in a public online database. The Ocean Genome Legacy operates under the sustainability principles established by the Convention on Biological Diversity.

Earth Charter

The Earth Charter is described on the Earth Charter Initiative Web site (www.earthcharter.org/) as a widely recognized, global consensus statement on ethics and values for a sustainable future. It is a declaration of fundamental values and principles considered necessary for building a sustainable global society in the 21st century. The Earth Charter was not successful in gaining support and adoption at the 1992 Rio Earth Summit, but two years later the time was right and in 1994, the movement to draft and implement the Earth Charter was restarted as a civil society initiative. The renewed efforts were the work of Maurice Strong, Chairman of the Earth Summit, with the help of the Earth Council, and Mikhail Gorbachev with the support of Green Cross. The government of The Netherlands also acted as a cornerstone of the renewed Earth Charter movement. Drafting of the text was overseen by the independent Earth Charter Commission, and the Commission continues to serve as the steward of the Earth Charter text. The Earth Charter was completed in March 2000 and launched at The Peace Palace in The Hague. Since its launch, the charter has been formally endorsed by thousands of organizations representing millions of people around the world. By 2005, the Earth Charter had become widely recognized as a global consensus statement on the meaning of sustainability, the challenge and vision of sustainable development, and the principles by which sustainable development is to be achieved. Beginning in 2007, national governments began to make even stronger, more formal commitments to the Earth Charter. Although not endorsed by any international governmental organization, the Earth Charter has a strong impact on global guidelines for sustainability and development goals. The Earth Charter continues to have increasing international impact as a source of inspiration for action, an educational framework, and as well as a reference document for the development of policy, legislation, and international standards and agreements.

The Earth Charter lives up to its name, having been created by a large global consultation process. The charter encourages the global community to help create a global partnership at a critical time in world development. The purpose of the Earth Charter is to inspire in all people a sense of interconnectedness and interdependence, as well as a shared responsibility for the human family and the larger living world. The Earth Charter urges environmental responsibility, peaceful coexistence, and respect for life, democracy, and justice to achieve its goals. The Earth Charter's ethical vision proposes that environmental protection, human rights, equitable human development, and peace are interdependent and indivisible. It provides a new framework for thinking about and addressing these issues. The Earth Charter is organized into 16 general headings, each covering a general principle.

These changed stories and beliefs may also help transcend some of the most powerful barriers to sustainability by enabling people to join in a common effort based on commonly held values, and revealing interconnectedness. These may help transcend limitations such as geopolitical boundaries and private property rights. These changed perspectives may also increase a sense of responsibility to and for future generations.

References

Chapple, Christopher Key, and Mary Evelyn Tucker, eds. 2000. *Hinduism and Ecology: The Intersection of Earth, Sky, and Water.* Cambridge, MA: Center for the Study of World Religions Harvard University Press.

Folz, Richard C. et al. eds. 2003. *Islam and Ecology: A Bestowed Trust.* Cambridge, MA: Center for the Study of World Religions Harvard University Press.

Mission, C. F. 2005. *Sharing God's Planet: A Christian Vision for a Sustainable Future.* London, UK: Church House Publishing.

Wangari Muta Maathai

Wangari Muta Maathai was born in Nyeri, Kenya, in 1940. She went on to become the first women in east and central Africa to earn a doctorate degree, but she did not stop there. She became the chair of the Department of Veterinary Anatomy and an associate professor for the University of Nairobi in 1976 and 1977. She was also the first woman to hold either of these positions in the region. During her veterinary positions, Wangari was active in the National Council of Women of Kenya and chaired the counsel from 1981–1987.

During her involvement with the National Council of Women, Wangari began the Green Belt Movement in 1976. The Green Belt Movement was a grassroots movement to get women's groups of Kenya to plant trees to improve the quality of life and the environment. Today, because of Wangari's vision of a better Africa, more than 30 million trees have been planted in women's gardens, church yards, and public lands; and more than 40 individuals from across Africa have established similar programs for their countries through Wangari's tutelage, including Tanzania, Uganda, Malawi, Lesotho, Ethiopia, and Zimbabwe. She and her Green Belt Movement have won numerous awards and accolades, most notably the Nobel Peace Prize in 2004.

Wangari is also internationally recognized for her dedication to the struggle for democracy, human rights, and environmental conservation. In 1998, she used her international recognition to launch a campaign called Jubilee 2000 Coalition. Sitting as co-chair, she campaigns globally for the forgiveness of unpayable debts backlogged against the poorest countries of Africa. Her work in protecting against land grabbing and rapacious allocation of forest land is also internationally recognized.

Wangari has most recently been elected to the Kenyan Parliament, with a 98 percent majority. The president of Kenya then appointed her as assistant minister for Environment, Natural Resources and Wildlife.

FIGURE 1.21 • Nobel Peace Prize laureate Wangari Maathai from Kenya, right, shakes hands with Norwegian Queen Sonja in Oslo City Hall, Norway, Friday, December 10, 2004. At center is King Harald V. AP Photo/ Bjorn Sigurdson, Pool.

COLLABORATIVE DECISION-MAKING PROCESSES

Collaborative environmental decision-making processes involve multiple stakeholders from the community, government, and industry to work together to solve an environmental problem. This is opposed to adversarial environmental decision-making methods that use courts and judicial processes to determine a winner and a loser. Other issues regarding the limitations of judicial approaches are that access to courts is restricted by income, politics, and legislation. This is especially true in environmental conflicts. There are few if any laws about sustainability, but there is a global consensus that sustainability is a major social goal. By restricting judicial accessibility, this forum makes incomplete and transient decisions that may actually impair sustainability. As environmental impacts accumulate and conflicts increase, new collaborative approaches are developing.

Collaboration in the use of natural resources offers benefits to sustainable approaches. It is being used more in watershed decisions and in some complex urban environmental issues. Collaboration allows for an evaluation of the full range of ecosystem services because of the multiple sources of information. It can increase the range of value perspectives on that use and the range of options for protecting and conserving that natural resource. By increasing information, value perspectives, and range of options, collaborative environmental approaches can also decrease uncertainty because the implications of various options are better known. In this way, uncertainty about the application of the precautionary principle to environmental decision making is decreased.

Collaborative environmental approaches may also be a way to span political boundaries that are not based on natural features. Because of this, collaboration may be confrontational, as stakeholders from other nations or cities are included. Collaborative environmental approaches may also become controversial, as formerly excluded groups are brought in as stakeholders. Eroded environmental resources, such as water, can present public health threats to formerly excluded groups. Issues of reparations and compensation for these damages may arise. This adds to the financial costs of collaborative environmental approaches, which are generally more expensive than traditional ways of environmental decision making. Collaborative environmental decisions may cost more because of the time and money of inclusion of new stakeholders and the need for capacity building in that system for complete and accurate ecosystem knowledge. Sustainability advocates, however, point out that those collaborative approaches may save money in the long run because they provide a fuller accounting of ecosystem services.

Collaborative approaches for sustainable environmental decision making also require that stakeholders are interested in the common good, not just self-interest. Cooperative, forthcoming, and engaging dialogues work better than adversarial, interest group, conflict-based negotiation in collaborative models. Alternative dispute resolution is not functional in collaborative sustainability decision making. The facilitation of multiple stakeholders, especially those with different capacities, is usually needed in the early stages of collaboration. The facilitator makes sure that all stakeholders have all the information, that they understand the information, and that they understand each other.

In the United States, New Jersey developed a collaborative environmental approach via their executive order on environmental justice. It has a mechanism for state citizens to file complaints that start a process that engages multiple stakeholders in the complaint, problem description, and problem solution. New Jersey is an older, industrialized state. It has a disproportionate number of known contaminated sites in African American, low income, and Latino urban areas. These sites have a number of public health issues such as asthma.

Once a complaint is filed from a community, the process engages many stakeholders. The Environmental Justice Task Force itself is composed of academic, public health, statewide environmental, civil rights, and public health organizations; large and small business and industry; municipal and county officials; and organized labor. At least one-third of the task force is from grassroots or faith-based community organizations. The task force can expand to include state agency expertise if the environmental complaint requires it. Although the executive order specifically requires that state agency constraints be considered, it also requires the involvement of senior executive staff of the agency if that expertise is needed. After the complaint is accepted, the task force develops a detailed action plan. This plan is designed to address the environmental, social, and economic factors that affect the health of the complaining community. Each step of the plan involves active and dynamic collaboration with state agencies, the community, and other stakeholders in the monitoring and enforcement of the plan. The task force itself strives to be completely accountable and inclusive of all stakeholders. Sometimes this even includes schools, and educational institutions are often ignored even though they contain vulnerable population with high exposures in contaminated areas. All types of information is developed; demographic, qualitative, and quantitative sources all count. The task force explicitly makes the complaining community part of the solution by asking it early on to access its own capacity to carry out solutions, and what it would need in order to do so. This collaborative aspect of the solution is in contrast to adversarial and judicial methods where the winner makes the loser either solve or mitigate a partial solution, or alternatively nothing environmental happens because the complainant failed to prove its case.

The New Jersey experience is useful in studying collaborative approaches to complex environmental problems and is promising for sustainability. By making the complainant part of the solution, environmental issues can be fully monitored. This will affect both public policies of the state, as well as personal patterns of private consumption. This process also brings in many people who fear involvement in environmental decision or complaints because they could lose their job, housing, or food sources. This brings in the type of information necessary for sustainable development, the exposed holes in the ecosystem, and is thus an example of the strength of collaboration—information gathering. The New Jersey Task Force also explicitly asks where a breakdown in communication between stakeholders occur. It keeps the collaboration strong to solve environmental and public health problems.

The New Jersey approach may be expensive under current methods of public policy evaluation, such as cost benefit. It is difficult, however, to place a value on new information, increased public health, and the new application of a holistic and ecological process. Part of the uncertainty in sustainable environmental decision-making approaches like collaboration is that past, incomplete processes were inexpensive because they were incomplete. The development of an ecosystem analysis that measures the past incompleteness as well as develops new measures and methods may have its only value in being sustainable.

The Common Sense Initiative

From 1995 to 1999, the Environmental Protection Agency (EPA) hosted an experiment in multistakeholder dialogue called the Common Sense Initiative. The EPA invited key stakeholders to participate in a series of dialogues to find alternatives to current governmental regulations that were "cheaper, cleaner, and smarter." Stakeholders included labor, environmental advocates, environmental justice groups, state and local governments, as well as the EPA. These dialogues brought together representatives of the three elements of sustainability in a unique process allowing for direct communications between people and organizations that rarely communicated in a constructive, face-to-face dialogue. All decisions were to be reached by consensus, although each sector committee could define the meaning of consensus for its processes.

Certain industrial sectors were selected to participate in this experiment based on their permitted emissions and other factors. Some of these sectors were primarily small businesses; others were dominated by large, multinational corporations. This sector-driven approach is a different approach to environmental protection from the traditional media-based approaches of the major statutes that the EPA enforces, such as air, water, and solid or hazardous waste deposited on land. Six industrial sectors were selected to participate in the dialogue: automobile manufacturing, computers and electronics, iron and steel, metal finishing, petroleum refining, and printing.

The six sector committees produced 43 different projects and led to other initiatives within the EPA including the Sector-Based Environmental Protection Action Plan and the Stakeholder Involvement Action Plan. Some projects focused on issues of pollution prevention and product lifecycle management. Others tackled issues related to monitoring of fugitive emissions and other information management challenges. One of the most comprehensive projects was designed by the printing sector. Its project, called PrintStep (Simplified Total Environmental Partnership), redefined permitting processes to tie pollution prevention to community involvement.

References

Hemmati, Minu et al. 2002. *Multi-Stakeholder Processes for Governance and Sustainability: Beyond Conflict and Deadlock.* London, UK: Earthscan.

Sabatier, Paul et al. eds. 2005. *Swimming Upstream: Collaborative Approaches to Watershed Management.* Cambridge, MA: MIT Press.

Urban Agriculture

The distance food and waste must travel can greatly increase the environmental footprint. The ecological footprint of a city is increased by reliance on nonrenewable energy sources required to transport food to markets in cities. One way to reduce the use of nonrenewable resources is by growing food and other crops closer to the city where it is consumed. This decreases the environmental footprint and increases self-reliance of city dwellers. Urban farming is the growing, processing, and distributing of food and other products through plant cultivation and animal husbandry in and near cities. Urban farming can be much more than simply growing vegetables. It can mean raising animals like cattle, dairy cows, chickens, pigs, horses, bees, fish, llamas, ducks, geese, and emus. Farm animals can create controversy in an urban context because of noise and waste. Many universities now offer courses in urban agriculture and community gardens.

Farming in cities can be a new use in urban residential and commercial areas. When permitted by land use laws, urban farmland may include community gardens, private window box and porch gardens, rooftop gardens, abandoned lots, and other sites not contaminated by prior industrial uses. Urban farmland may also include tillable land called greenbelts around cities. Competition for greenbelt land comes from residential development of suburban homes and large retail or strip shopping areas. This is one of the controversies involved in land use and sustainability.

Urban farmers include nonprofit food providers to poor people, churches, schools, and others who grow food for their own consumption.

Garden projects are a common form of urban agriculture. There are different emphasis among programs. Some help empower indigenous people and increase a knowledge of community nutrition. Others are commonly associated with missions or other organizations that aid homeless and poor people in urban areas.

Some examples of garden projects follow.

- In Los Angeles, California, food is being grown on urban walls, in this case organically. The project will provide organic greens to people who live there, soften and green hardscapes of concrete and steel, and diminish the heat sink effects and global warming (see Urban Farming, www.urbanfarming.org/foodchain2.htm).

- In Santa Cruz, California, the Homeless Garden Project provides employment to homeless people growing organic food on a 2.5-acre garden. This project "seeks to provide a place of safety, beauty, and work opportunities for people who were homeless."

• In Aboriginal Community Kitchen Gardens at UBC Farm, Vancouver, BC, "members of the Musqueam First Nation, have grown vegetables on the farm site for their community kitchen project since 2002. With an interest in expanding the potential benefits of this community nutrition project, the farm initiated a new pilot program in 2005. In collaboration with 17 different agencies working on Vancouver's Downtown Eastside (DTES), a plot of land on the farm is dedicated towards the DTES Aboriginal Community Kitchen Garden Project. Clients at community kitchen and food bank projects not only have the opportunity to get fresh produce, but also come out to the farm to grow it. Students, in turn, have a great opportunity for on-campus community service learning." (see City Farmer News, www.cityfarmer.info/urban-aboriginal-community-the-garden-project-at-ubc-farm/).

• Lansing, Michigan Food bank "administers 18 community gardens in the Lansing area, by securing land and arranging for plowing, and rototilling. All gardener participants receive supplies including seeds, plants, canning supplies, and training, which enable them to grow and preserve their own fresh vegetables. More than 440 families receive supplies and technical assistance for growing their own gardens. The Garden Project also organizes crews of volunteers to harvest surplus fruits and vegetables from area farms. The produce is then distributed to food pantries, human service organizations, and to residents of low-income housing, particularly to senior citizens. Volunteer harvesting efforts provide more than 200,000 pounds of fresh fruits and vegetables each season."

When the cost of transportation of goods becomes too expensive for consumers to bear, they will be forced back into localized goods and products. This is a countermeasure to the effects of globalization moving employment in certain manufacturing and agricultural sectors to distant locations in order to save labor costs. When transportation costs exceed labor costs as a cost of production, localized supplies will regain their market saliency.

Agriculture in particular will need to adapt to the high cost of transportation. Some foresee that this will force localization of foods, meaning increasing the amount of agricultural production near urban areas. This change has been severely compromised by urban sprawl and the environmental impacts of population growth in urban centers taking both land and water resources away from local agriculture.

One idea for localization of food resources in urban areas is vertical farming. Traditional farming is done on the surface of the earth at the

Heifer Learning Center at Overlook Farm

Heifer International's Learning Center is an urban agricultural experience designed to educate people around the issue of global hunger and poverty. It is located in Rutland, Massachusetts, about 60 miles outside of Boston. As the project's Web page says, "The programs offer the extraordinary opportunity for people to experience some of the challenges of global hunger and poverty—and come away with a re-energized determination to be part of the solution."

Education opportunities include tours, Global Village immersion experiences, Service-Learning programs and specifically designed tours for groups. According to their Web page, these programs are designed to be "experiential, hands-on, interactive and fun, Overlook is also a working example of the sustainable agriculture Heifer supports around the world."

There is a heavy reliance on volunteers, and this educational aspect of urban sustainability is highly developed at Heifer Learning Centers. Volunteers are an integral part of leadership and programming as well as relied on to work in the gardens and care for the livestock.

Overlook Farm is a 270-acre working farm. The farm demonstrates sustainable agricultural practices and the ways in which they are applied in real situations. Some of the urban agricultural practices they teach are intensive rotational animal husbandry, organic farming approaches, and sustainable forestry practices. Every year more than 14,000 people come to Overlook Farm to volunteer in tours, work groups, educational programs, or to attend special events. At any given time, more than 100 volunteers are working the farm. Money from program fees, special events, and the sale of farm products goes to run the farm.

Reference
Heifer International, www.heifer.org/site/c.edJRKQNiFiG/b.201558/#.

level of the topsoil. Among the challenges of traditionally configured farming is the need to acquire and manage large tracts of land usually outside of the market area for which the food is grown. Acquiring large tracts of arable land is expensive and may destroy wetlands or impinge species habitat. In addition, managing such tracts may require substantial investment in machinery and infrastructure such as irrigation or drainage, and farm equipment that uses fossil fuel for its energy. Vertical farming is more similar to greenhouse growing operations, done in multistory buildings resembling skyscrapers. It is compact in its impact on the actual surface acreage preserving wetlands and other important land uses including species habitat. In addition, locating these sky farm buildings in and near urban markets will greatly reduce the cost of transportation and thus the cost of food. At least five designs for sustainable sky farms are being considered by architects and city planners. Sky farms would decrease the ecological footprint of a city because they would decrease the transportation and energy costs of moving food to the city and moving wastes from the city. One of the first sky farm designs is the vertical farm project by Chris Jacobs with Dr. Dickson Despommier. In this design tall, self-contained towers produce more energy, water (via condensation/purification), and food than their occupants consume. Some architects are developing designs that fit into already existing buildings to make them into sky farms. In Seattle, Washington, the architectural design company, Mithun, has designed a sky farm specifically for downtown Seattle. Downtown Seattle is a mixture of residential- and commercial-use properties. The goal of the design is to make the building self-sufficient. It relies on nature for water, recycles gray water, uses solar energy for power, and builds in farming spaces.

Architect Gordon Graff has designed a sky farm for the city of Toronto, Canada. His designs include 48 stories and large amounts of space for raising crops. Because Canada has a short growing season, the ecological footprint of cities like Toronto can be large because food has to be transported into the city and waste taken out of the city. Mr. Graff's design would greatly decrease the ecological footprint because his sky farm would grow food right there and use "waste" to grow more food.

The city of Las Vegas, Nevada is also considering building a sky farm. Designs now call for a 30-story building with a sky farm that grows food for the large number of tourists. Although sky farm designs are still in the design phase, with none beginning construction, many of the ideas about energy, food consumption, waste recovery, and ecological footprint are specifically aimed at sustainability. ***See also* Volume 1, Chapter 4: Agriculture.**

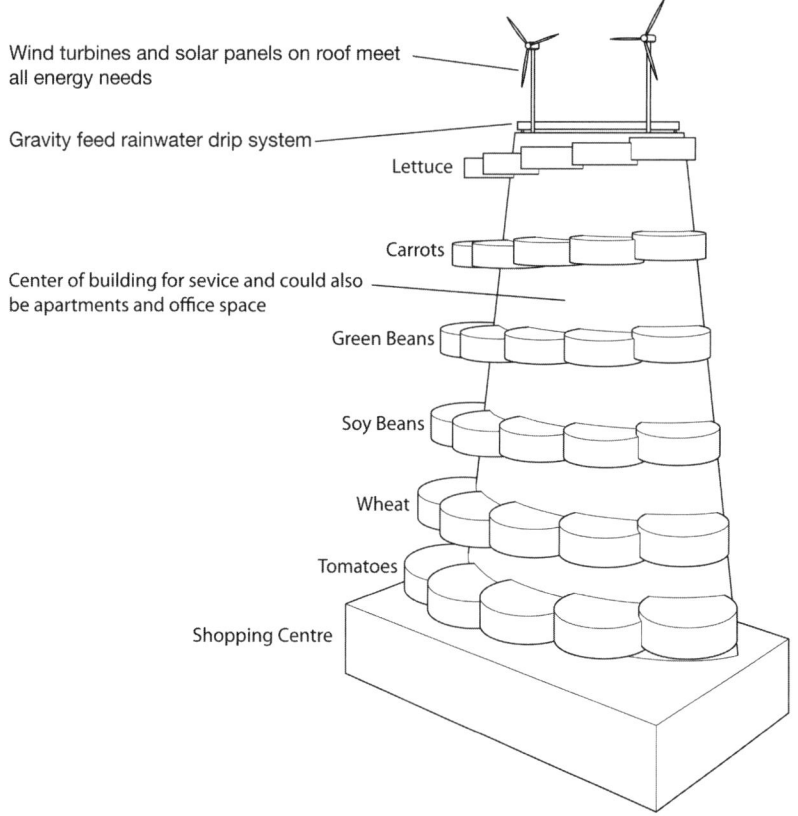

FIGURE 1.22 • Sky farms or vertical farms grow crops in urban skyscrapers rather than traditional fields that use more horizontal square footage. Illustrator: Jeff Dixon.

Ocean Mining for Freshwater

Nymphea-Water is a French undersea research and engineering corporation that examines mining freshwater in the ocean before it becomes diluted by seawater. Basically a form of underwater prospecting, the company relies on local fisherman, infrared cameras to detect temperature differences., and geologic maps to find fresh water springs under the sea. Limestone cliffs are hosts to many underground springs. When rainwater falls on the limestone it reacts with the calcium in the limestone to form an acid that cuts into the limestone and forces its way down. It collects in empty spaces under the limestone to form underground reservoirs. In the last ice age water was captured by glaciers. When the glaciers began to melt about 21,000 years ago, they resupplied these underground caverns. Ocean levels rose, and when the pressure of the seas was higher than the outflow pressure from the now underwater rivers, the underground rivers and springs slowed or stopped their flows.

Underwater mining is a dangerous proposition for humans. Nitrogen gas poisonings occur from rapid ascent, and oxygen deprivation occurs from descents that are too long. Technological advancements in undersea robotics and personal submarines assist the new miners.

Reference

Community Food Security Coalition. 2003. Urban Agriculture and Community Food Security in the United States: Farming from the City Center to the Urban Fringe, http://www.foodsecurity.org/PrimerCFSCUAC.pdf.

Capitalism and Sustainability

Capitalism is a term used to describe free market-based economies both at an ideal or theoretical level and as a current operating global political economy. Sometimes the free market is also called the "private" sector. This implies a lack of control by government, although in actuality it is the state that allows many businesses to exist. It is one of the most powerful and prevailing value structures in the world today, as well as in recent history. Many feel that it is poised for a period of robust growth, with far-reaching environmental consequences. In terms of sustainability, there are many fertile grounds for controversy because most versions of capitalism run counter to preservation of environment for future generations. As former Vice President and Nobel Prize winner Al gore noted:

> capitalism's recent triumph over communism should lead those of us who believe in it to do more than merely indulge in self congratulation. We should instead recognized that the victory of the West—precisely because it means the rest of the world is now more likely to adopt our system—imposes upon us a new and even deeper obligation to address the shortcomings of the capitalist economics as it is now practiced.
>
> The hard truth is that our economic system is partially blind. It 'sees' some things and not others. It carefully measures and keeps track of the value of those things most important to buyers and sellers. . . . But its intricate calculations often completely ignore

the value of other things that are harder to buy and sell: fresh water, clean air, the beauty of the mountains, the rich diversity of life in the forest . . . the partial blindness of our current economic system is the single most powerful force behind what seem to be irrational decisions about the global environment.

Al Gore, *Earth in Balance: Ecology and the Human Spirit.*
Boston: Houghton Mifflin, 1992, pp. 182–83.

Capitalism as a basic concept usually incorporates several characteristics. Private producers of goods and services hire workers. They produce the good or service with the sole intent, and a legal responsibility, to make a profit. Any activity that reduces profits is suspect. Economic development becomes a strong value of the people and of the government. The free market development of capitalism becomes developed and supported by government. Corporations are creatures of the state in that they are allowed to form according to the laws passed by the government. The basic form of corporate structure allows for corporations to invest money without individual liability for the debts they incur. This decreases the risk of many financial transactions and allows for entrepreneurs, or risk takers, to seek new markets and develop new technologies.

Capitalism as a complicated concept involves multinational corporations with budgets bigger than many nations. Sometimes these nations have natural resources, some that could be irreparably damaged and could be necessary for future generations. A large enough multinational corporation in search of profits might disregard the potential harms to a community or even a nation. For example, a large multinational corporation that mines gold in a very dry desert region might disregard the harm to the ecosystem, surrounding community, and even national economy if the profits from mining were sufficient. Gold mining takes water and discharges arsenic in the process. In a very poor country, capitalism exerted by multinationals may prevent fundamental principles of sustainability from being operationalized. The application of the precautionary principle, where the risks of irreversible damage to natural systems is first assessed, is expensive and time consuming.

Capitalism as a complicated concept includes materialistic values, a strong public policy emphasis on economic development, and higher rewards for the strongest competitors. Materialistic values are expressed when identity and status become functions of owning and purchasing. Consuming goods and services for the value of consuming, or social status, and without regard to environmental impacts is materialism. As one scholar on growth noted:

capitalism itself has ans wered the demands that inspired 19th century socialism . . . But the attainment of these goals has only brought deeper sources of social unease –manipulation by marketers, obsessive materialism, environmental degradation, endemic alienation. . . . we have the freedom to consume instead of the freedom to find our place in the world.

Clive Hamilton, 2004, pp. 112–13.

National, state, and local governments around the world develop many public policies around economic development. Although they are not all capitalistic, they all seek to improve the quality of life of at least present-day humans. The role of government and the type of government intervention into the free or private market are greatly affected by values of capitalism. Governments can respond to strong political pressure from legitimate industry groups and carve out special exceptions or subsidies for industries that show promise of high profits under the name of economic development. Governments can also offer transactional security and privacy for business organizations such as multinational corporations, partnerships, subsidiaries, banks, and other lending institutions.

References

Gore, Al. 1992. *Earth in the Balance: Ecology and the Human Spirit.* Boston: Houghton Mifflin.

Hamilton, Clive.2004. *Growth Fetish.* London: Pluto Press.

Government Invention in the Free Market Under Capitalism

Government intervention in most forms interferes with perceived profits. However, any government regulation that provides for government contracts, grants, or subsidies is sought after by industry. Governmental intervention in this manner is considered misleading by environmentalists because this type of government intervention distorts the true environmental impact of the good or service. Government subsidies, along with other avenues other than government power and influence, may exacerbate industrial practices that could jeopardize some of the basic principles of sustainability. Sustainability proponents argue that these government subsidies should go to industries that are trying or supporting sustainability.

At the local level, many towns and villages are seeking industry to improve employment as part of their economic development package. In a capitalistic approach to land and private property (see Volume 1, Chapter 3: Government and UN Involvement, Local and Regional Governmental Activities), the local government will seek to streamline regulatory processes designed to protect the environment and public welfare, abate property and utility taxes for a set period of time, and clean up and give away industrial sites When regulatory requirements are overlooked at the local level, they can also be overlooked at the state and national level. There are many projects in the United States with significant impacts on the environment that do not file any type of environmental impact assessment. These regulatory requirements are often minimal and act to protect the environment. Without them, overdevelopment, pollution, and unsustainable economic development occur. The problem is that mispricing goods and services by ignoring environmental impacts of production and consumption causes a misuse of the natural resource base. This upsets sustainability proponents because part of sustainability principles is that systems of life on which life depends are kept in as good a state for future generations.

As capitalism and democracy gain world popularity, some aspects of capitalism may be reaching their environmental limitations. Some environmental amenities have no present value because there is no value to future lives. As countries, states, and cities grapple with growing populations wanting to consume more under capitalism and democracy develop public polices, they will also face environmental limitations. This includes the climate changes to be wrought by global warming, natural resource scarcity, and food shortages. Some current predictions have the greatest climate impact on the equator and surrounding tropical regions. Population and food supply rest on a precarious balance in many nations of this region. If drought and desertification continue along with population growth, then environmental limitations will be much more severe for future generations.

All these features work against sustainable development. Materialistic values in a capitalistic free market that does not place a transactional value on natural resources threaten to consume resources and damage ecosystems. In one way of stating it, the goose that lays the golden egg has been killed; that is, these systems were destroyed before knowledge of the damage was politically powerful enough to find expression in government. That is one reason why sustainability proponents advocate for the precautionary principle. The high demand on natural resources of a growing population in a new global economy burgeoning with capitalism and free markets has not been planned, contains many unknown dimensions, and causes many conflicts and controversies around values. Agenda 21 addressed sustainability in agriculture in the context of population growth estimates:

> By the year 2025, 83 percent of the expected global population of 8.5 billion will be living in developing countries. Yet the capacity of available resources and technologies to satisfy the demands of this growing population for food and other agricultural commodities remain uncertain. Agriculture has to meet this challenge, mainly by increasing production on land already in use and by avoiding further encroachment on land that is only marginally suitable for cultivation.

> Paragraph 14:1. United Nations Conference on Environment and Development, Agenda 21, U.N. Document A/CONF.151.26 (1992).

Government facilitation of capitalism as a value and a public policy undergirding economic development also works against sustainability because it supports unsustainable but politically powerful industrial development. This is particularly true in agribusiness. The small, self-reliant family farm is a romantic idea from the past in most industrialized countries. Subsidies of tobacco, cotton, milk, rice, and other crops ignore their environmental impact, but respond to political constituencies at the local, state, national, and sometimes international level. Other nations may choose to subsidize the crops of their own powerful political constituencies. All of these types of government that developed economic policies to attract and enable free markets to flourish may come at an environmental price. *See also* **Volume 2, Chapter 2: Market–Based Strategies.**

Reference

Paragraph 14:1. United Nations Conference on Environment and Development, Agenda 21, U.N. Document A/CONF.151.26 (1992). www.un.org/esa/dsd/resources/res_publcorepubli.shtml.

Global Capitalism and Its Relationship to Global Environmental Conditions

The planet has seen a vast expansion of economic activity, even more than population growth. Many environmental writers today believe this rapid and unchecked economic expansion is the main cause of environmental degradation. The world economy is better organized and now operates on a global scale. Environmentalists, sustainability advocates, and others are concerned that a large population on the verge of economic growth, primarily via some type of "capitalism," will increase the rate of environmental degradation to further decrease the potential for sustainability. As capitalism becomes less an economic theory but more a political cause, it causes a failure to recognize all the nonmarket costs of production. One of the main costs that are ignored under capitalism is environmental impacts. These can occur in the use of raw materials, in the industrial processes of production; in the shipment, storage, and use of the product; and in the disposal of the product. These ignored environmental impacts can accumulate and damage ecosystems. They can take the form of air pollution, toxic and hazardous waste sites, and polluted water. Environmental degradation and public health decline are also part of nonmarket costs. Capitalistic political economies, such as the United States, also support many production activities that produce serious and unaccounted environmental effects. Some examples of these are mining, logging, and ranching. Some industries are virtually environmentally unregulated and industry self-reports much of the information. Modern environmental advocacy groups and most approaches to environmental public policy are inadequate to the global challenges presented by current versions of capitalism. Capitalism practiced this way has socialized the harms and risks from speculation and exploitation and privatized the gains from these activities to corporations and their shareholders. Several recent propositions have to do with the bank and home mortgage lender bailouts. "Banks are too big to fail, and homeowners are too small to bail." ***See also* Volume 2, Chapter 2: Globalization.**

Reference

Parr, Adrian. 2009. *Hijacking Sustainability*. Cambridge, MA: MIT Press.

Harnessing Economic Forces for Sustainability

Emerging versions of capitalism are developing new strategies for approaching the demands of society for a sustainable community. The free market is considered important for sustainability because of the innovation and technological advancement that is characteristic of emerging businesses and industry. There are several themes to these emerging theories. Most call for greatly increased efficiency from resource use. Increased efficiency can be in the form of reuse, recycle, and reduce use of natural resources, especially those that are nonrenewable. Industrial ecology focuses on the elimination of "wastes" in any production cycles, finding other uses for them and sometimes protecting profit. Many see an economy that moves from the production of goods to the production of services as the necessary step toward sustainability.

Why Is Growth Necessary Under Capitalism?

The global economy has grown roughly 5 percent a year, the U.S. economy about 3.5 percent. It is a fast rate of growth, primarily fueled by political capitalism and via multinational corporations. In 2019, the world economy could double in size at a 5 percent growth rate.

Economic growth is a strong value that undergirds capitalism. The function of growth underscores high and fast growth of products. Economic growth is not limited to capitalism. Other political economies also value economic growth, as noted by one scholar:

> Communism aspired to become the universal creed of the twentieth century, but a more flexible and seductive religion succeeded where communism failed: the quest for economic growth. Capitalists, nationalists—indeed almost everyone, communists included—worshiped the same alter because economic growth disguised a multitude of sins. Indonesians and Japanese tolerated endless corruption as long as economic growth lasted. Russians and eastern Europeans put up with clumsy surveillance states. Americans and Brazilians accepted vast social inequalities. Social, moral, and ecological ills were sustained in the interest of economic growth; indeed, adherents to the faith proposed that only

more growth could resolve such ills. Economic growth became the indispensable ideology of the state nearly everywhere."

> J. R. McNeill, 2000, *Something New Under the Sun: An Environmental History of the Twentieth Century World,* New York: W. W. Norton, 2000, pp. 334–336.

Growth is traditionally measure by gross domestic product (GDP), although this measure has become more controversial as global concern about the environment has increased. This is discussed in more detail in volume 2.

There are other ways to measure growth, but few capture how economic growth affects the environment. One reason for this is that measures of economic growth are dependent on political boundaries of nations and states, not environmental regions. Economic growth can occur by sacrificing the environment for short-term gains that are not sustainable. In a 1996 Report on Human Development, the United Nations Development Program examined the economic growth of nations and discussed some types of growth under traditional measurements. They found five kinds of growth buried within traditional economic measures of growth:

1. Futureless growth in which the present generation consumes resources on which future generations depend

2. Jobless growth in which the overall economy grows but jobs do not grow.

3. Selfish growth in which only the top income levels benefit from growth.

4. Growth in loss of cultural identity in which the economic growth causes people's cultural identity to fade.

5. Politically oppressive growth in which economic growth occurs with a decrease in political empowerment of the people.

Because of the lack of inclusion of these factors and because of the lack of inclusion of environmental impacts, some are directly challenging economic growth indicators and economic growth. Some aspects of this controversy deal with whether the loss of raw materials will cause a lack of growth and whether a society should grow. This has developed into an area known as ecological economics.

The main definition of ecological economics comes from a 2004 book on the topic called *Ecological Economics.* Here the authors directly challenge economic growth because of the lack of inclusion of environmental impacts.

> More contentious is the call by ecological economics for an end to growth. We define growth as an increase in throughput, which is the flow of natural resources from the environment, through the economy,

and back to the environment as waste. It is a quantitative increase in the physical dimensions of the economy and/or the waste stream produced by the economy. This kind of growth, of course, cannot continue indefinitely, as the Earth and its resources are not infinite. While growth must end, this in no way implies and end to development, which we define as qualitative change, realization of potential, evolution toward an improved, but not larger, structure or system . . .

Where conventional economics espouses growth forever, ecological economics envisions a steady—state economy at optimal level. . . . The difference could not be more basic, more elementary, or more irreconcilable.

Daly and Farley, *Ecological Economics,* pp. 6, 23.

Another controversial measure of economic growth is the Index of Sustainable Economic Welfare (ISEW). This measure starts with a baseline of national private consumption expenditures. It then adjusts that figure for inequalities in their distribution. Then it adds nonmarket contributions to welfare. These would be volunteer efforts, childcare, and housework. It also subtracts so-called defensive expenditures such as fire and police protection, pollution control and abatement, and the loss of value of natural resources and environmental assets. The ISEW has begun to morph into another indicator that is applied with essentially the same dynamics. That indicator is called the genuine progress indicator. Other controversial indicators are emerging. They are controversial primarily because they are new and challenge the GDP indicator. One of these is the weighted index for social progress developed by Richard Estes at the University of Pennsylvania. This indicator and others like it attempt to measure social and environmental conditions. Another emerging indicator is the happy planet index developed by the New Economics Foundation in the United Kingdom. It tries to capture human life satisfaction and health by multiplying the life satisfaction of a nation by its life expectancy. It then divides this number by the ecological footprint of that country. Low environmental impacts with long, happy lives yield a high score.

References

Daly, Herman E., and Joshua Farley. 2004. *Ecological Economics.* Washington, DC: Island Press.

Elkins, Paul. 2000. *Economic Growth and Environmental Sustainability.* London, UK: Routledge.

Hamilton, Clive. 2004. *Growth Fetish.* London, UK: Pluto Press.

McNeill, J. R. 2000. *Something New Under the Sun: An Environmental History of the Twentieth Century World.* New York: W. W. Norton.

The Economic and Environmental Consequences of Growth

What is the record of economic development, spurred by burgeoning capitalism and rapid globalization? For sustainability advocates of rapid and radical social changes, this is a key question. The answer for many in the world is cumulative environmental degradation and a widening division between rich and poor. *See also* **Volume 2, Chapter 4: The Economic and Environmental Consequences of Growth.**

The poor of industrialized nations bear the disproportionate environmental burden of this growth. It is difficult to persuade nations, communities, and peoples suffering a low quality of life to act in a sustainable manner for the sake of the planet. The record of ignored environmental impacts is just emerging, and most aspects of it are controversial. Multinational industries reliant on natural resources are exploiting the economic condition of nations suffering a low quality of life, especially in the areas of mining and petrochemical production. Environmentalists charge that they are moving fast because the world community is enacting tougher and more enforceable environmental restrictions on their activities. Industry responds that it is simply reacting to market demand. Environmentalists are now much more aware of global environmental actions. The strength of the environmental movement is their shared observations of nature because in most places in the world, the only observers are the people who live there. The strength of industry, however, is its rapid growth and political power. Over half the largest economies in the world are multinational corporations, not nations.

Economic growth has increased the quality of life and negatively impacted the environment. As measured by GDP, the world economy increased by a factor of 14 from 1890 to 1990. In the same period, the global population increased by a factor of 4, water use increased by a factor of 9, sulfur dioxide emissions increased by a factor of 13, energy use increased by a factor of 16, carbon dioxide emissions increased by a factor of 17, and the marine fish catch increased by a factor of 35. In the first century of capitalism, economic growth increased rapidly and environmental impacts continued to be ignored. Of concern to modern-day sustainability advocates are the unknown cumulative, synergistic, and ecological impacts of such rampant growth because these types of impacts could irreversibly affect natural systems of land, air, and water on which all future life depends. If economic growth continues to increase, the concern is that even more environmental damage could occur.

One dynamic that many feel portends rapid economic growth is the mechanism of global financial markets, which often drive the financing for other markets. High profit growth is the primary value. One way this is observed is growth in market capitalization and the price paid for its stock. The expected rate of profit growth is very important. Losses often cause market value to decrease. In a competitive capitalistic market, most major financing industries need to finance highly profitable industries and corporations to ensure they meet their profit expectations. As finance markets are now global and operate 24 hours a day seven days a week in many places, the ability to quickly and irreparably exploit a weakness in a nation's environmental laws to overuse a natural resource in a unsustainable and profitable manner is much greater. The dynamic of finance markets lead many to believe that some nations could pursue a path of unsustainable economic growth as free markets grow into rapidly increasing populations. All cumulative environmental impacts to date, and what are now small environmental impacts, may increase in scale as population and capitalism increase. The threat of irreversible damage to systems of life like air and fresh water could increase dramatically.

Reference

Gabel, Medard, and Henry Bruner. 2003. *Global Inc.: An Atlas of the Multinational Corporation.* New York: New Press, pp. 2–3.

RECENT EVIDENCE OF ECONOMIC GROWTH AND ENVIRONMENTAL IMPACT

There are many challenges to economic growth as capitalism of any type in many parts of the world. Other nations may have different values, and measure gross national happiness as in Bhutan. Many locations do not have addresses of personal or commercial property. Land ownership and recordation may be nonexistent in written form. Some value structures that do not allow the forfeiture of a home for payment of a debt, perhaps because the property is communally owned, or that landownership is unclear. In these types of nations, many lending institutions are hesitant to finance projects because without the security of clear ownership to property as collateral, they could lose profit. Other nations do not have convertible currency, or currency valuable or stable enough to be traded for other currencies on the world market. The relationship of the government to its people is also very different in many places. These three factors'—lack of identifiable place, nonconvertible currency, and shifting and unknown nation/people relationships—pose as much an obstacle for capitalism as they do for sustainability.

A serious challenge to economic growth, and therefore capitalism in many cases, is concern about the environmental impacts. Although the overall trends discussed in the preceding paragraphs describe events over a century, the question becomes more pointed when it addresses whether environmental regulations mitigated environmental impacts enough to be a basis for sustainability. Right now, most researchers answer no. By examining economic growth and environmental impacts since the rise of environmental regulatory protocols, there are still economic growth and environmental impacts, although they are lower than the growth of the world economy. Environmental impacts are still getting worse, but at a slower rate. From 1980 to 2005, according to the World Resources Institute (www.earthtrends.wri.org) and the Worldwatch Institute (www.worldwatch.org/node/1066/print), the gross world product increased 46 percent. During the same time period, paper production and fish catch increased 41 percent, meat consumption increased 37 percent, energy use increased 23 percent, fossil fuel use increased 20 percent, the world population increased 18 percent, nitrogen oxide emissions increased 18 percent, freshwater use increased 16 percent, carbon dioxide emissions increased 16 percent, fertilizer use increased 10 percent, and sulfur dioxide emissions increased 9 percent. All these measures are fraught with some degree of controversy. They do show that with environmental regulation it is possible to make a difference, but they also show that the difference is not sufficient.

A great concern to sustainability advocates is the characteristic of capitalism to fuel exponential growth. Each year's successive outputs and profits are invested to increase the rate and amount of production. So, too, does waste and environmental degradation exponentially increase. Unlike economic growth that has no theoretical limit, however, the global ecology has definite limits to its capacity to regenerate itself. Once these capacities are breached, ecosystems may reach a tipping point of no return. This generally means that that major ecosystems in the land and water are irreparably damaged beyond the point of their resiliency.

Market advocates claim that the market is self-correcting. They point to technology as capable of mitigating environmental impacts because it can reduce the amount of natural resources consumed and produce new products that impact less on the environment.

Right now, this is not happening fast enough for sustainability advocates. The latest research from five large West European and U.S. research centers examined these very issues. They concluded:

> Industrial economies are becoming more efficient in their use of materials, but waste generation continues to increase. . . . Even as a decoupling between economic growth and resource throughput occurred on a per capita and per unit GDP basis, overall resource use and waste flows into the environment continued to grow. We found no evidence of an absolute reduction in resource throughput. One half to three quarters of annual resource inputs to industrial economies are returned to the environment as wastes within a year.

James Gustave Speth, 2008, p. 56.

Reference

Speth, James Gustave. 2008. *The Bridge at the Edge of the World: Capitalism, the Environment, and Crossing from Crisis to Sustainability.* New Haven, CT: Yale University Press.

Classic Capitalism and the Environment: Implications for Sustainability

As political economies embrace both capitalism and democracy, many search for the original basis for capitalistic theories. Given their robust growth, sustainability advocates also examine these works for environmental or ecological content. Some of the foundational works of classical capitalism recognized the threat to the environment. One scholar noted in 1944:

> To allow the market mechanism to be the sole director of the fate of human beings and their natural environment . . . would result in the demolition of society. Nature would be reduced to its elements, neighborhoods destroyed and landscapes defiled, rivers polluted, military safety jeopardized, the power to produce food and raw materials destroyed."

Karl Polanyi, 1944, p. 73.

A fundamental critique of classical capitalism is the preference for the present over the future. The future is always unknown. The value of science to capitalism is in its ability to predict the future within a given probability and range of error. Sustainability expressly values future lives, at least to the extent present lives are valued. Future generations cannot be part of current supply and demand trends, or markets. Their number and value are unknown, controversial, and may not be quantifiable. When governments try to begin regulatory standards based on sustainability concerns for future generations, they often face this issue of the value of future lives. Because governments are often regulating corporations and their environmental behavior, and because corporations value profits over unknown future lives, corporations often resist any government regulation of their environmental behavior for purposes of sustainability. For example, one aspect of sustainability is the precautionary principle. Here the environmental impact of a proposed activity is assessed to determine if irreversible damage to a natural system could occur. If so, there is usually an exploration of what mitigative actions could and should be taken. All this environmental assessment takes time and ties up capital in what is perceived as a high-risk financial proposition. The emphasis of the present over the future under capitalism also resists expenditures of time and money that does not contribute to profit.

The question of a future value of a human life is a question that is answered in a different context right now. In legal cases of negligence, the value of a wrongful death is computed by juries.

References

Adger, W. Neil et al., eds. 2006. *Fairness in Adaptation to Climate Change*. Cambridge, MA: MIT Press.

DiMento, Joseph F. C., and Pamela Doughman, eds. 2007. *Climate Change: What It Means for Us, Our Children, and Our Grandchildren*. Cambridge, MA: MIT Press.

Emanuel, Kerry. 2007. *What We Know about Climate Change*. Cambridge, MA: MIT Press.

Parry, M. L., O. F. Canziani, et al., eds. 2007. *Climate Change 2007: Impacts, Adaptation and Vulnerability. Contribution of Working Group II to the Fourth Assessment Report of the Intergovernmental Panel on Climate Change*. Cambridge, UK: Cambridge University Press.

Polanyi, Karl. 1944. *The Great Transformation.* Boston: Beacon Press.

Volk, Tyler. 2008. *CO2 Rising: The World's Greatest Environmental Challenge.* Cambridge, MA: MIT Press.

EDUCATION ON THE ENVIRONMENT FOR SUSTAINABILITY

Academia and Sustainability

Educational institutions have been engaged in planning and thinking about sustainability for at least two decades. Most of their efforts have been aimed at achieving sustainability in their operations much like any other business organization. These efforts include lowering environmental impacts from their physical plants through reductions in toxins and waste in areas including grounds, housekeeping, and food services. In addition, they have focused on achieving energy efficiency in the construction and renovation of buildings and the use of energy for light and heat. Some also encourage these types of measures in their suppliers and contractors. Some have also tried to incorporate long-term vision for sustainability in their financial operations including endowment investments.

Curricular commitments to sustainability as a teachable subject remain more elusive. One reason for this curricular disconnection may lie in the way that higher education has organized its disciplinary structure in contemporary colleges and universities. The modern university is organized around distinctions between the arts, sciences, and professions. Liberal arts are usually based on the classical areas of learning including history, philosophy, rhetoric and politics, and mathematics. Sciences are often distinguished by the branches of scientific learning that developed during the enlightenment period, including biology, chemistry, geology, and physics. These distinctions are the basis for the degree-granting and teaching structures within most universities. Sometimes, these distinctions are compared to agricultural silos in the sense that they keep various fields separate and coherent, but they prevent mixing that might result in new combinations.

Sustainability crosses the frontiers of multiple disciplines, from social sciences, to physical sciences, and into areas like law, business, and engineering, which are often associated with professional training.

References

Allen-Gil, Susan et al. 2008. *Addressing Global Environmental Security through Innovative Educational Curricula.* New York: Springer.

Leal Filho, Walter, and Mario Salomone, eds. 2006. *Innovative Approaches to Education for Sustainable Development.* New York: Peter Lang.

Rappaport, Ann, and Sarah Hammond Creighton. 2007. *Degrees That Matter: Climate Change and the University.* Cambridge, MA: MIT Press.

Operations and Curriculum

Many institutions have moved their operational conduct toward sustainability. Campuses often operate as businesses do, including their grounds keeping, housekeeping, food services, building, and housing functions. In addition, campuses resemble businesses in their relationships with suppliers and contractors. They may also have substantial sums of money to manage in the form of endowments from benefactors. Many campuses have found that reducing their environmental footprint in these operations is attractive to their students and makes practical sense in both short- and long-term financial efficiencies. Many institutions of higher education have moved toward an emphasis on sustainability in their operations.

Moving toward sustainability in terms of what is taught in the curriculum has been significantly more difficult to achieve. This is likely due to the organization of the disciplines within the contemporary university system, as well as their systems of rewards within the different disciplines. In addition to challenging traditional differences between disciplines, sustainability as a field of study is new and embraces uncertainty in a way that is uncommon to the traditional fields of learning.

Reference

Thomashow, Mitchell. 2003. *Bringing the Biosphere Home: Learning How to Perceive Global Environmental Change.* Cambridge, MA: MIT Press.

Uncertainty Squared: The Challenge of Sustainability

Education at all levels focuses on teaching what we know and what we are reasonably sure about based on facts and inferences. At the higher and more abstract levels, education attempts to teach ways to think about what we are less sure of when facts are uncertain and inferences are attenuated. That is one level of uncertainty. When faced with uncertainty at this level, education responds with philosophies and values based on discussions that can be charted and certified by degree-granting programs. A different experience of uncertainty occurs when we do not know what we do not know. At that level of uncertainty, traditional education has had difficulty in conferring degrees and pedigrees. Many of the issues involved in the study of sustainability involve this second level of uncertainty. Beyond conflicting and competing values invoked by topics such as those noted in the controversies section of each volume, the complexity of factual data that comes from each place studied makes education a moving target, crossing many academic terrains, and occasionally requiring the honest admission of ignorance and humility.

The ultimate challenge for education in this situation is how to teach about complexity and ontological and epistemological ignorance. The ultimate challenge for a society struggling with issues of sustainability is how to make decisions under these circumstances.

Some meta-disciplines have long struggled with these levels of uncertainty and the need for decision making. One such discipline is law. Law confronts the need for certainty in facts with a procedural acknowledgment that fact is socially constructed. In this discipline, fact-finding is done by a jury of social peers recognizing that their conclusions are reflective of the community's own cultural perspectives, not necessarily a scientific enterprise. They are often instructed in how to resolve uncertainty in terms of a spectrum of probabilities ranging from a mere preponderance to a moral certainty beyond reasonable doubt.

When asked to make policy recommendations, scientists may experience the discomfort of approaching disciplinary limits in the form of uncertainty and the political consequences of uncertainty. Theories such as post-normal science attempt to place traditional sciences within a matrix of other knowledge such as community experience for the purpose of escaping the paralysis of judgment that uncertainty can impose within traditional limits.

These multidisciplinary, meta-disciplinary approaches to the problems of uncertainty are not yet comfortably housed within the architecture of contemporary educational structures, challenging educational institutions to come up with new structures.

References

Azeiteiro, Ulisses, Fernando Goncalves, Walter, Leal Filho, Fernando Morgado, and Mario Pererira, eds. 2004. *World Trends in Environmental Education.* New York: UNIPUB.

Corcoran, Peter Blaze, and Arfen E. J. Wals. 2007. *Higher Education and the Challenge of Sustainability: Problematics, Promise, and Practice.* New York: Springer.

Creating a Constituency for Sustainability: Public Outreach

There can be no lasting change in a society without an educational policy and program to support that change. This was the operating theory that has guided many developing countries struggling to develop in a postcolonial, postindustrial world. The great economies of China and India developed after colonialism and industrialism relied on their abilities to educate their population to perform complex, nonmenial work.

The task facing developed countries in the struggle to achieve global sustainability is to do the same at a time when commitment to public education financed by local property taxes has weakened. In developed democracies, education has not kept pace with the public need and interest in sustainability as an area of study and coherent policy development even as the support for public education has dwindled.

Currently, public outreach about anything sustainable is minimal. In the United States, many still perceive environmental regulation as "liberal" and as an impediment to job creation and economic growth. The lack of public outreach causes a lack of public knowledge about sustainability. This in turn causes a lack of public engagement with

Arizona State University and the School of Sustainability

In 2007, Arizona State University (ASU) established a new school dedicated to the study of sustainability. A retreat hosted by ASU in Yucatan Peninsula brought together academic visionaries and leaders in the field of sustainability. They were asked what a school of sustainability should teach, and why their own institutions were not presently teaching these things. The result was the establishment of a unique multidisciplinary school granting undergraduate and graduate multidisciplinary degrees. The curriculum is largely based on these pioneering discussions about what we need to learn about what we do not yet know.

election of candidates who promise to protect the environment, and this in turn stifles the development of research and public policy in the area of sustainability.

In some pioneering parts of higher education, however, new research is emerging about sustainable approaches in industry, government, and communities. Neighborhoods in Seattle, Washington are starting a grassroots effort for sustainability. The city of Seattle has formed an Office of Sustainability and Environment. Its goal is to collaborate with other city agencies, nonprofit groups, business organizations, and neighborhood organizations to provide outreach to the public about actions that advance sustainability. It has emphasized environmental actions and results. The mayor wants to make Seattle the green building capitol of the United States. One way to meet this goal is to reduce greenhouse gas emissions, and he has agreed to try to reduce greenhouse gas levels to 7 percent below 1990 levels by 2012 and 80 percent of 1990 levels by 2050. He wants to increase energy efficiency in new and existing buildings by 20 percent. He has developed an extensive public outreach program to meet this goal. He has also developed other programs that save Seattle's trees, create a community greenhouse gas inventory, create a community carbon footprint, and develop an implementation plan.

Neighborhoods are a particular focus in this example of a public outreach effort. Seattle has created a Sustainable Urban Neighborhoods Initiative. This initiative works closely with neighborhoods to evaluate their strengths and weaknesses in terms of sustainability. It organizes and facilitates meetings among neighborhoods, universities, and city agencies. It does on-the-ground research about environmental conditions in that neighborhood. It also helps increase the capacity of neighborhoods to understand what the research and indicators mean to the residents.

All across the United States, many large urban areas are grappling with the costs of ignoring centuries of environmental impacts and hoping that public outreach for sustainability will improve the economic, ecological, and equitable aspects of the quality of life.

Reference

Tilbury, Daniella, and David Wortman. 2004. *Engaging People in Sustainability*. Gland, Switzerland: International Union for Conservation of Nature and Natural Resources.

ENVIRONMENTAL BASIS OF SUSTAINABILITY AND CAPITALISM

The environmental basis of sustainability relates to exceeding the carrying capacity of current ecosystems. When the systems of nature on which present and future life depends become irreparably damaged, there is a global moral consensus that this is wrong. This can be seen in watersheds that are so polluted that nothing grows in them or in global warming and climate changes that threaten present and future life. Population growth, knowledge of past and present environmental impacts, and increase in environmental impacts all contribute to a consensus that we can and do exceed the environmental capacity of our planet to sustain us. There are still controversies about the science, the actual impact, and who should bear the cost of environmental restoration. Implicit values are made explicit when they reach a gridlock. Gridlock is reached when none of the traditional decision-making functions of society work and conflicts continue to fester. Many societies are in a state of environmental gridlock, and implicit values around capitalism are being made explicit.

Economics is a social science that studies the supply and demand of commodities. It is an implicit value that sustainability makes explicit. Other social sciences, such as psychology and sociology, also have value assumptions. All social sciences, like all human activities, cannot exceed the boundaries of nature for long. If something is not easily commoditized, then economics struggles to make it fit into its paradigm of supply and demand. Capitalism takes it a step further, although the term has been politically manipulated to serve dominant cultures. Capitalism relies on free markets, motivated and knowledgeable entrepreneurs, and the free flow of credit without discrimination based on race, religion, social class, gender, ethnicity, or political belief.

The natural capitalism movement uses the language of economics and the values of capitalism to try to create a sustainable and restorative economy. It explicitly recognizes the limits of growth as environmentally imposed, and that growth through perpetual resource depletion is not sustainable. The restorative economy restores eroded ecosystems, and protects and preserves working ecosystems. It makes promises of radical technological innovation, newfound prosperity, meaningful work, and low-risk security.

One of the trends and indicators of a restorative economy is that corporations and other profit-motivated organizations begin to evaluate their motivations and behavior from the perspective of their environmental impact over time. Profit as a proxy for social good is rejected in a restorative economy. The ability of a business or community to integrate ecological systems in its means of production and distribution is the measure of its success. This would require radical changes in current taxing schemes and business laws. The restorative economy is also indicated by manufacturer responsibility for their products, as well as for their wastes. Another indicator is true cost environmental accounting for the production of goods and provision of services. Transactions are readily apparent to all stakeholders, and their impact on the environment is known. Externalities under economics and pollution under capitalism are part of every transaction and are not the basis for competition. Another trend in the restorative economy is that goods and services are created locally or regionally. Nationally or internationally produced goods have a greater environmental impact and are not as transparent as locally produced goods.

Another set of trends of a restorative economy is that distant sources of capital are not used. Distant sources of capital may originate in sources that excessively deplete natural systems. Restorative economies also respect their labor forces and use humans more than automation to make their products. They recognize that the function of work is not just corporate profit, but also social and community identity. The products created are built for long-term use, and when they cease to function they will not harm future generations. Planned obsolescence of a product is not considered part of restorative economy. Consumers are educated to understand and appreciate environmental value in products and services. Most proponents of an ecologically restorative society believe it will come from an infusion of environmental justice, from the grassroots up.

References

Bell, Simon, and Stephen Morse. 2008. *Sustainability Indicators: Measuring the Immeasurable?* London, UK: Earthscan.

Bevington, Douglas. 2009. *The Rebirth of Environmentalism: Grassroots Activism from the Spotted Owl to the Polar Bear.* Washington, DC: Island Press.

Cunningham, Storm. 2002. *The Restoration Economy.* San Francisco: Berret-Koehler Publishers.

Hawken, Paul. 1993. *The Ecology of Commerce.* New York: Collins Business.

INFORMATION

Information is critical to sustainable decision making in the future; information about ecological and environmental conditions, as well as human health and economic information. Basic data are missing, important information is not shared, and information sources are ignored or corrupted. Communities are invaluable repositories of information about the history and human uses of land and resources. Community knowledge must be systematically collected and considered in sustainable decision making. Scientific knowledge has been significantly distorted or discarded by some governments. This corruption of data impedes sustainable public policymaking.

The Need for Adequate Baseline Information about the Environment

Information about environmental and ecological conditions is collected by various levels of government that do not share it with each other. Creating forums in which such information is shared is important to integrated decision making. For example, land use and developmental decisions should consider human health and environmental data. *See also* **Volume 1, Chapter 4: Information.**

Reference

Doremus, Holly. "Data Gaps in Natural Resource Management: Sniffing for Leaks Along the Information Pipeline." *Indiana Law Journal* 83 (2008): 407–463.

Glicksman, Robert L. "Bridging Data Gaps Through Modeling and Evaluation of Surrogates." *Indiana Law Journal* 83 (2008): 465–525.

Mitchell, Ronald B., William C. Clark, David W. Cash, and Nancy M. 2006. *Global Environmental Assessments: Information and Influence.* Cambridge, MA: MIT Press.

RISK ANALYSIS: COMMONLY HELD RISK PERCEPTION

Risk is a way of determining the allocation of resources. Dangers that are real require some plan of action to ensure security. But how much to invest in avoidance depends greatly on the perception of risk associated with that danger, including its probability of occurrence and the severity of consequences if it does occur. In addition, investment in avoidance varies greatly according to the social and political arrangements underlying contingencies for dealing with danger.

Ecosystem risks connected to rapid and widespread change in our environment and supporting webs of life are difficult to identify and plan for because of their attenuated nature. We do not know enough about the facts of such complicated and interrelated dynamics to fully predict their impact. We do not even know what we do not know (epistemological ignorance). In this situation, it is hard for those enjoying the benefits of their current position to appreciate why there is a need for change. It is also hard for those who are suffering the hardships of their current position to appreciate the need to develop and grow in a different way from the past.

This combination of ignorance, coupled with a contemporary unequal distribution of ecological benefits and burdens, has contributed to a stalemate of political wills in developed and developing countries. Developed countries are faced with the challenge of changing climate and energy resources still enjoy advantages that historical policies conferred. The need for change of a radical nature may still seem remote. Developing countries faced with a tide of environmental refugees (people whose lives have been disrupted by environmental degradation) and natural disasters are under tremendous pressure to develop resources for the benefit of their suffering populations. The common platform necessary for these countries to agree on an agenda of sustainability is a commonly held perception of the risks of ecological collapses and their interrelatedness.

Uncertainty in Facts and Values

One of the most difficult aspects of reaching decisions in any complex context is the problem of uncertainty about facts. There are different levels of factual uncertainty. Some facts can be known with effort; some cannot be assessed even with effort because of the failure of record keeping or lack of technological skill. Some facts cannot be known because we do not know what we do not know. This is called epistemological uncertainty, and it is the most radical level of factual uncertainty. Epistemological uncertainty is characteristic of most ecological crises.

One of the greatest areas of uncertainty relates to values. Individuals and cultures vary greatly in their values regarding whether the environment should be preserved for future generations. In many privileged parts of a given society, the values are not examined because it is simply presumed they are right. Whenever it is claimed something is "value free" or "value neutral," it means that the values are the implicit traditional values that one accepts without question. Values can change radically in the same person. If the impact of those values affects them personally, they may have a different view than a value that only affects others.

As knowledge about the environment and environmental degradation increases, more values will clash. Values of high levels of economic growth and short-term profits for the wealthy will clash with values of those who prefer no growth and the redistribution of wealth. Nations, religions, large business organizations, and other social institutions that form around certain value clusters may also clash. The globalization of science, however, makes it go beyond serving just one value set, as in a nation.

Scientific expertise, expertise in values clarification, and post–normal scientific expertise in ecological risk assessment can become a trend to hone values in a way that supports the development of sustainable development policies. The globalization of science that occurred with global warming predictions and intensive scientific study of climate changes offers the possibility of transcending limited science and moving into unlimited and dynamic science. *See also* **Volume 1, Chapter 2: Waste, Pollution, and Toxic Substances.**

Reference

Hlavinek, Petr et al. 2009. *Risk Management of Water Supply and Sanitation Systems.* New York: Springer.

SCIENCE FOR POLICY

Scientists are often challenged in this context as not being independent, or not contributing meaningful knowledge to the process fast enough to serve the values of sustainability and environmental protection. To many, scientists quickly defend and serve the values of profit and economic growth.

The role of science in protecting and advancing the values of environmental protection and sustainability is well illustrated in the case of global warming. The United States is the largest greenhouse gas emitter per capita and in total pounds. Greenhouse gases are now indisputably a part of global warming. From a global standpoint, the United States needs to reduce its greenhouse gas emissions to reduce climatic entries. Because scientists rely on the principle of scientific uncertainty, the U.S. government relied on this principle to refuse to develop a meaningful global warming policy. The United States backed away from signing the Kyoto Protocol, as well as President Bush's campaign promises to regulate carbon dioxide emissions from electric utilities. At the same time, the United States was embracing the precautionary principle as enunciated in the 1992 United Nation Framework Convention on Climate Change. That language states that:

> The Parties should take precautionary measures to anticipate, prevent or minimize the causes of climate change and mitigate its adverse effects. Where there are threats of serious or irreversible damage, lack of full scientific certainty should not be used as a reason for postponing such measures, taking into account that policies and measures to deal with climate change should be cost-effective so as to ensure global benefits at the lowest possible cost.

> United Nations Framework Convention on Climate Change, May 29, 1992, UN Doc. A:AC.237/18 (1992)

The United States therefore committed itself to not using the principle of scientific uncertainty to prevent reduction of greenhouse gases, but it did so nevertheless. It is likely that the emission of some greenhouse houses caused by the United States increased under the Bush administration if the wartime greenhouse gas emissions from the U.S. military presence in Iraq and Afghanistan are included. The U.S. military is the largest consumer of petrochemical products and used them in military engagements in these and other countries. These emissions technically come from the geographical location of another country, but are caused by U.S. activities. Science and scientifically developed technological advancements were heavily incorporated in these military activities. At the same time science and scientifically developed technological advancements were not heavily incorporated in environmental protection or sustainable development.

Eventually, the Bush administration did propose a greenhouse gas emission program that relied on voluntary acts as long as they did not harm the U.S. economy. The U.S. economy at that time was among the most robust in the world. In addition to being the largest green-house gas emitter, the United States is also one of the nations that could afford to reduce its emissions. Many greenhouse gas emissions in the United States could be reduced quickly and cheaply with the engage-ment of scientists and those who apply science in everyday life. Some of these very simple policies include citizen education programs about global warming, research and support for alternative energy programs, emission limitations for power plants, building and land use codes that require high energy efficiency, environmental labeling requirements for energy sources, transparent environmental transactions of industries with high greenhouse gas emissions, and a greater customer choice of energy suppliers. This short list of basic programs would not cost any more than current energy expenditures but would probably save U.S. consumers money.

The failure of the United States to support science in the area of greenhouse gas emissions despite promises to the contrary has deepened international distrust of U.S. leaders and scientists. The United States continued to make large profits in nonrenewable energy sources, such as gas and oil, while continuing to emit the most greenhouse gases. Greenhouse gases now cause climate changes that can have devastat-ing effects around the globe, from flooding, rising ocean levels, and drought to species extinctions and famine. Some nations now propose that the United States be the waste site for their own energy expan-sion programs. For example, in the late 1990s, China considered using the United States as the repository for its nuclear waste as it developed its nuclear energy program. Its reasoning was that because the United States has benefited the most and polluted the planet the most, it is now time for other nations that reach the level of prosperity of the United States to do so. Because other nations have to bear the burden of U.S. pollution, it is now the turn of the United States to bear its proportion-ate share of global climate change.

International tension around scientific uncertainty and U.S. com-mitment to global warming and climate change run deep. It will be a significant challenge for future leaders in the United States scientific community to move toward sustainable development.

References

Adger, W. Neil et al. eds. 2006. *Fairness in Adaptation to Climate Change.* Cambridge, MA: MIT Press.

Egan, Michael. 2009. *Barry Commoner and the Science of Survival.* Cambridge, MA: MIT Press.

Hess, David J. 2007. *Alternative Pathways in Science and Industry: Activism, Innovation, and the Environment in an Era of Globalization.* Cambridge, MA: MIT Press.

Ohshita, Stephanie B. "The Scientific and International Context for Climate Change Initiatives." *University of San Francisco Law Review* 42 (Summer 2007): 1–37.

United Nations Framework Convention on Climate Change, May 29, 1992, UN Doc. A:AC.237/18 (1992) available online at unfccc.int/resource/docs/convkp/con-veng.pdf.

Appendix A: Portal Web Sites

Al Gore www.algore.com

Aldo Leopold, The Wilderness Society, www.wilderness.org/aboutus/leopold.cfm

All Birds Barcoding Initiative, www.barcodingbirds.org

Army Environmental Policy Institute, www.aepi.army.mil

Association for the Advancement of Sustainability in Higher Education, www.aashe.org/

Barcode of Life Initiative, www.dnabarcodes.org

Berkeley Sustainability Assessment sustainability.berkeley.edu/assessment

Blueprint for a Green Economy Submission to the Shadow Cabinet, Quality of Life Policy Group Chairman, Rt Hon John Gummer MP, Vice-Chairman, Zac Goldsmith, September 2007 (p. 391) (available on line at www.qualityoflifechallenge.com)

Botanic Gardens Conservation International, www.bgci.org/plant_search.php

Campus Report Card www.nwf.org/campusEcology/campusreportcard.cfm

Campus Sustainability Snapshot, www.njheps.org/assessment/guide.htm

Canadian Barcode of Life Network, www.bolnet.ca

City Farmer News, www.cityfarmer.info/urban-aboriginal-community-the-garden-project-at-ubc-farm/

Community Food Security Coalition, www.foodsecurity.org

Decision Framework For Special Reports, Methodology Reports and Technical Papers, www.ipcc.ch/pdf/ipcc-principles/revd-decision-framework-for-special-reports.pdf

The Designers Accord, www.designersaccord.org

Earth Charter www.earthcharter.org/

Ecological Footprint www.redefiningprogress.org/footprint/

Electronic Product Environmental Electronics Assessment Tool, www.epeat.net

Federal Register, www.gpoaccess.gov/fr/

Fish Barcode of Life Initiative, www.fishbol.org

Fort Carson Sustainability and Environmental Management System Plan, www.fedcenter.gov/_kd/Items/actions.cfm?action=Show&item_id=10217&destination=ShowItem

Fourth Climate Action Report to the UN Framework Convention on Climate Change, www.state.gov/g/oes/rls/rpts/car/

Global Reporting Initiative www.globalreporting.org/AboutGRI/WhoWeAre/

Gro Harlem Brundtland at www.af-info.or.jp/eng/honor/hot/enr-brundtland.html

Heifer International, www.heifer.org/site/c.edJRKQNiFiG/b.201558/#

Homeless Garden Project, www.homelessgardenproject.org

In Memoriam www.rachelcarson.org

Initial Report of the United States to the United Nations Committee on the Elimination of Racial Discrimination (2000) available at www.state.gov/www/global/human_rights/cerd_report

International Property Rights Index, www.InternationalPropertyRightsIndex.org

IPCC, Assessment Priorities, www.ipcc.ch/pdf/ipcc-principles/revd-decision-framework-for-special-reports.pdf

Klamath Basin Index Page, www.andykerr.net/KlamathBasin/KlamathBasinPT.htm

Limitations on US landfill capacity soils.usda.gov/survey/geography/hurricane/index.html

Millennium Declaration,www.un.org/millennium/declaration/ares552e.htm

Millennium Seed Bank Project online at www.kew.org/msbp/visit/index.htm

Our Common Future, available on line at www. un-documents.net/wced-ocf.htm

Profiles of the 2004 Blue Planet Prize Recipients, Dr. Gro Harlem Brundtland at www.af-info. or.jp/eng/honor/hot/enr-brundtland.html

Rachel Carson Institute, Commemorating Rachel Carson, www.chatham.edu/RCI/aboutrc.html

Rachel Carson, www.rachelcarson.org

Redefining Progress, www.redefiningprogress.org

Report of the World Commission on Environment and Development, www.un.org/documents/ga/ res/42/ares42–187.htm

Report of the World Commission on Environment and Development: Our Common Future www. un-documents.net/wced-ocf.htm

Silent Spring Introduction by Vice President Al Gore, www.uneco.org/ssalgoreintro.html

The Jane Goodall Institute, www.janegoodall.org/ jane/default.asp

Tidal Power Plant, www.technologystudent.com/ energy1/tidal3.htm

Toxics Release Inventory by Zipcode, www.scorecard. org.htm

United Nations Environment Programme, www. unep.org

Urban Farming, www.urbanfarming.org/foodchain 2.htm

US EPA, Brownfields and Land Revitalization website www.epa.gov/swerosps/bf/index.html

US EPA, Brownfields www.epa.gov/swerosps/bf/ index.html

USEPA, Index of Watershed Indicators, at epa.gov/iwi/

US EPA, Watersheds Approach, www.epa.gov/ owow/watershed/approach.html

Vandana Shiva, www.southendpress.org/authors/17

Wangari Maatha. www.greenbeltmovement. org/w.php?id=3, 7/22/2008

"Who was Aldo Leopold" at www.naturenet.com/ alnc/aldo.html

World Resources Institute www.earthtrends.wri.org

Worldwatch Institute (www.worldwatch.org/ node/1066/print

Appendix B: Rio Declaration on Environment and Development

RIO DECLARATION ON ENVIRONMENT AND DEVELOPMENT

The United Nations Conference on Environment and Development,

The United Nations met at Rio de Janeiro from 3 to 14 June 1992, reaffirming the Declaration of the United Nations Conference on the Human Environment, adopted at Stockholm on 16 June 1972, and seeking to build upon it,

With the goal of establishing a new and equitable global partnership through the creation of new levels of cooperation among States, key sectors of societies and people,

Working towards international agreements which respect the interests of all and protect the integrity of the global environmental and developmental system,

Recognizing the integral and interdependent nature of the Earth, our home,

Proclaims that:

PRINCIPLE 1

Human beings are at the centre of concerns for sustainable development. They are entitled to a healthy and productive life in harmony with nature.

PRINCIPLE 2

States have, in accordance with the Charter of the United Nations and the principles of international law, the sovereign right to exploit their own resources pursuant to their own environmental and developmental policies, and the responsibility to ensure that activities within their jurisdiction or control do not cause damage to the environment of other States or of areas beyond the limits of national jurisdiction.

PRINCIPLE 3

The right to development must be fulfilled so as to equitably meet developmental and environmental needs of present and future generations.

PRINCIPLE 4

In order to achieve sustainable development, environmental protection shall constitute an integral part of the development process and cannot be considered in isolation from it.

PRINCIPLE 5

All States and all people shall cooperate in the essential task of eradicating poverty as an indispensable requirement for sustainable development, in order to decrease the disparities in standards of living and better meet the needs of the majority of the people of the world.

PRINCIPLE 6

The special situation and needs of developing countries, particularly the least developed and those most environmentally vulnerable, shall be given special priority. International actions in the field of environment and development should also address the interests and needs of all countries.

PRINCIPLE 7

States shall cooperate in a spirit of global partnership to conserve, protect and restore the health and integrity of the Earth's ecosystem. In view of the different contributions to global environmental degradation, States have common but differentiated responsibilities. The developed countries acknowledge the responsibility that they bear in the international pursuit to sustainable development in view of the pressures their societies place on the global environment and of the technologies and financial resources they command.

PRINCIPLE 8

To achieve sustainable development and a higher quality of life for all people, States should reduce and eliminate unsustainable patterns of production and consumption and promote appropriate demographic policies.

PRINCIPLE 9

States should cooperate to strengthen endogenous capacity-building for sustainable development by improving scientific understanding through exchanges of scientific and technological knowledge, and by enhancing the development, adaptation, diffusion and transfer of technologies, including new and innovative technologies.

PRINCIPLE 10

Environmental issues are best handled with participation of all concerned citizens, at the relevant level. At the national level, each individual shall have appropriate access to information concerning the environment that is held by public authorities, including information on hazardous materials and activities in their communities, and the opportunity to participate in decision-making processes. States shall facilitate and encourage public awareness and participation by making information widely available. Effective access to judicial and administrative proceedings, including redress and remedy, shall be provided.

PRINCIPLE 11

States shall enact effective environmental legislation. Environmental standards, management objectives and priorities should reflect the environmental and development context to which they apply. Standards applied by some countries may be inappropriate and of unwarranted economic and social cost to other countries, in particular developing countries.

PRINCIPLE 12

States should cooperate to promote a supportive and open international economic system that would lead to economic growth and sustainable development in all countries, to better address the problems of environmental degradation. Trade policy measures for environmental purposes should not constitute a means of arbitrary or unjustifiable discrimination or a disguised restriction on international trade. Unilateral actions to deal with environmental challenges outside the jurisdiction of the importing country should be avoided. Environmental measures addressing transboundary or global environmental problems should, as far as possible, be based on an international consensus.

PRINCIPLE 13

States shall develop national law regarding liability and compensation for the victims of pollution and other environmental damage. States shall also cooperate in an expeditious and more determined manner to develop further international law regarding liability and compensation for adverse effects of environmental damage caused by activities within their jurisdiction or control to areas beyond their jurisdiction.

PRINCIPLE 14

States should effectively cooperate to discourage or prevent the relocation and transfer to other States of any activities and substances that cause severe environmental degradation or are found to be harmful to human health.

Principle 15

In order to protect the environment, the precautionary approach shall be widely applied by States according to their capabilities. Where there are threats of serious or irreversible damage, lack of full scientific certainty shall not be used as a reason for postponing cost-effective measures to prevent environmental degradation.

Principle 16

National authorities should endeavor to promote the internalization of environmental costs and the use of economic instruments, taking into account the approach that the polluter should, in principle, bear the cost of pollution, with due regard to the public interest and without distorting international trade and investment.

Principle 17

Environmental impact assessment, as a national instrument, shall be undertaken for proposed activities that are likely to have a significant adverse impact on the environment and are subject to a decision of a competent national authority.

Principle 18

States shall immediately notify other States of any natural disasters or other emergencies that are likely to produce sudden harmful effects on the environment of those States. Every effort shall be made by the international community to help States so afflicted.

Principle 19

States shall provide prior and timely notification and relevant information to potentially affected States on activities that may have a significant adverse transboundary environmental effect and shall consult with those States at an early stage and in good faith.

Principle 20

Women have a vital role in environmental management and development. Their full participation is therefore essential to achieve sustainable development.

Principle 21

The creativity, ideals and courage of the youth of the world should be mobilized to forge a global partnership in order to achieve sustainable development and ensure a better future for all.

Principle 22

Indigenous people and their communities and other local communities have a vital role in environmental management and development because of their knowledge and traditional practices. States should recognize and duly support their identity, culture and interests and enable their effective participation in the achievement of sustainable development.

Principle 23

The environment and natural resources of people under oppression, domination and occupation shall be protected.

Principle 24

Warfare is inherently destructive of sustainable development. States shall therefore respect international law providing protection for the environment in times of armed conflict and cooperate in its further development, as necessary.

Principle 25

Peace, development and environmental protection are interdependent and indivisible.

Principle 26

States shall resolve all their environmental disputes peacefully and by appropriate means in accordance with the Charter of the United Nations.

Principle 27

States and people shall cooperate in good faith and in a spirit of partnership in the fulfillment of the principles embodied in this Declaration and in the further development of international law in the field of sustainable development.

Source: Report of the United Nations Conference on the Human Environment, Stockholm, June 5–16, 1972.

APPENDIX C: AGENDA 21

Agenda 21 is a comprehensive plan of action to be taken globally, nationally, and locally by organizations of the United Nations System, Governments, and Major Groups in every area in which humans impact on the environment. The Preamble of the plan appears below. The complete document may be viewed online at the Web site of the UN Department of Economic and Social Affairs' Division for Sustainable Development (http://www.un.org/esa/dsd/).

1.1. Humanity stands at a defining moment in history. We are confronted with a perpetuation of disparities between and within nations, a worsening of poverty, hunger, ill health and illiteracy, and the continuing deterioration of the ecosystems on which we depend for our well-being. However, integration of environment and development concerns and greater attention to them will lead to the fulfilment of basic needs, improved living standards for all, better protected and managed ecosystems and a safer, more prosperous future. No nation can achieve this on its own; but together we can—in a global partnership for sustainable development.

1.2. This global partnership must build on the premises of General Assembly resolution 44/228 of 22 December 1989, which was adopted when the nations of the world called for the United Nations Conference on Environment and Development, and on the acceptance of the need to take a balanced and integrated approach to environment and development questions.

1.3. Agenda 21 addresses the pressing problems of today and also aims at preparing the world for the challenges of the next century. It reflects a global consensus and political commitment at the highest level on development and environment cooperation. Its successful implementation is first and foremost the responsibility of Governments. National strategies, plans, policies and processes are crucial in achieving this. International cooperation should support and supplement such national efforts. In this context, the United Nations system has a key role to play. Other international, regional and subregional organizations are also called upon to contribute to this effort. The broadest public participation and the active involvement of the non-governmental organizations and other groups should also be encouraged.

1.4. The developmental and environmental objectives of Agenda 21 will require a substantial flow of new and additional financial resources to developing countries, in order to cover the incremental costs for the actions they have to undertake to deal with global environmental problems and to accelerate sustainable development. Financial resources are also required for strengthening the capacity of international institutions for the implementation of Agenda 21. An indicative order-of-magnitude assessment of costs is included in each of the programme areas. This assessment will need to be examined and refined by the relevant implementing agencies and organizations.

1.5. In the implementation of the relevant programme areas identified in Agenda 21, special attention should be given to the particular circumstances facing the economies in transition. It must also be recognized that these countries are facing unprecedented challenges in transforming their economies, in some cases in the midst of considerable social and political tension.

1.6. The programme areas that constitute Agenda 21 are described in terms of the basis for action, objectives, activities and means of implementation. Agenda 21 is a dynamic programme. It will be carried out by the various actors according to the different situations, capacities and priorities of countries and regions in full respect of all the principles contained in the Rio Declaration on Environment and Development. It could evolve over time in the light of changing needs and circumstances. This process marks the beginning of a new global partnership for sustainable development.

APPENDIX D: THE MILLENNIUM ECOSYSTEM ASSESSMENT

Living Beyond Our Means: Natural Assets and Human Well-being (Statement of the Millennium Ecosystem Assessment Board)

Key Messages

- Everyone in the world depends on nature and ecosystem services to provide the conditions for a decent, healthy, and secure life.

- Humans have made unprecedented changes to ecosystems in recent decades to meet growing demands for food, fresh water, fiber, and energy.

- These changes have helped to improve the lives of billions, but at the same time they weakened nature's ability to deliver other key services such as purification of air and water, protection from disasters, and the provision of medicines.

- Among the outstanding problems identified by this assessment are the dire state of many of the world's fish stocks; the intense vulnerability of the 2 billion people living in dry regions to the loss of ecosystem services, including water supply; and the growing threat to ecosystems from climate change and nutrient pollution.

- Human activities have taken the planet to the edge of a massive wave of species extinctions, further threatening our own well-being.

- The loss of services derived from ecosystems is a significant barrier to the achievement of the Millennium Development Goals to reduce poverty, hunger, and disease.

- The pressures on ecosystems will increase globally in coming decades unless human attitudes and actions change.

- Measures to conserve natural resources are more likely to succeed if local communities are given ownership of them, share the benefits, and are involved in decisions.

- Even today's technology and knowledge can reduce considerably the human impact on ecosystems.

They are unlikely to be deployed fully, however, until ecosystem services cease to be perceived as free and limitless, and their full value is taken into account.

- Better protection of natural assets will require coordinated efforts across all sections of governments, businesses, and international institutions. The productivity of ecosystems depends on policy choices on investment, trade, subsidy, taxation, and regulation, among others.

BIBLIOGRAPHY

Ackerman, Frank, and Lisa Heinzerling. 2004. *Priceless: On Knowing the Price of Everything and the Value of Nothing.* New York: New Press.

Adger, W. Neil et al. eds. 2006. *Fairness in Adaptation to Climate Change.* Cambridge, MA: MIT Press.

Aksoy, Ataman M., and John C. Beghin. 2005. *Global Agricultural Trade and Developing Countries.* Washington, DC: The World Bank.

Allen, Jenny Smart. 2006. *Permaculture Design.* UK: New Holland Publishers.

Allen-Gil, Susan et al. 2008. *Addressing Global Environmental Security through Innovative Educational Curricula.* New York: Springer.

Azeiteiro, Ulisses, Fernando Goncalves, Walter Leal Filho, Fernando Morgado, and Mario Pereira, eds. 2004. *World Trends in Environmental Education.* New York: UNIPUB.

Baber, Walter F., and Robert V. Bartlett. 2005. *Deliberative Environmental Politics: Democracy and Ecological Rationality.* Cambridge, MA: MIT Press.

Bachmann, Peter, Michael Kohl, and Risto Paivinen. 1998. *Assessment of Biodiversity for Improved Forest Planning.* New York: Springer.

Bacon, Christopher M. et al., eds. 2008. *Confronting the Coffee Crisis: Fair Trade, Sustainable Livelihoods and Ecosystems in Mexico and Central America.* Cambridge, MA: MIT Press.

Bailey, Gilbert Ellis. 2008. *Vertical Farming.* Glacier National Park, MT: Kessinger.

Beer, Tom, and Alik Ismail-Zadeh. 2003. *Risk Science and Sustainability: Science for the Reduction of Risk and Sustainable Development of Society.* New York: Springer.

Bell, Simon, and Stephen Morse. 2008. *Sustainability Indicators: Measuring the Immeasurable?* London, UK: Earthscan.

Benfield, F. Kaid et al., 2001. *Solving Sprawl: Models of Smart Growth in Communities across America.* Washington, DC: Island Press.

Binley, Dan, and Oleg Menyailo. 2004. *Tree Species Effects on Soils: Implications for Global Change.* New York: Springer.

Blewitt, John. 2008. *Community Development, Empowerment and Sustainable Development.* Devon, UK: Green Books.

Buckingham, Susan, and Kate Theobald. 2003. *Local Environmental Sustainability.* Boca Raton, FL: CRC.

Bunnell, Gene. 2002. *Making Places Special: Stories of Real Places Made Better by Planning.* Chicago: Planners Press, American Planning Association.

Calow, Peter. 1998. *Handbook of Environmental Risk Assessment and Management.* Lanham, MD: Government Institutes.

Capra, Fritjof. 1996. *The Web of Life: A New Scientific Understanding of Living Systems.* New York: Anchor Books.

Carson, Rachel. 1962. *Silent Spring.* Boston: Houghton Mifflin.

Chapple, Christopher Key, and Mary Evelyn Tucker, eds. 2000. *Hinduism and Ecology: The Intersection of Earth, Sky, and Water.* Cambridge, MA: Center for the Study of World Religions Harvard University Press.

Clapp, Jennifer, and Peter Dauvergne. 2005. *Paths to a Green World: The Political Economy of the Global Environment.* Cambridge, MA: MIT Press.

Clark, William C. et al. 2006. *Linking Knowledge with Action for Sustainable Development: The Role of Program Management.* Washington, DC: National Academies Press.

Collin, Robert W. 2006. *The Environmental Protection Agency: Cleaning Up America's Act.* Westport, CT: Greenwood Press.

Collin, Robert W. 2008. *Battleground: Environment.* Westport, CT: Greenwood Press.

Committee on Environmental Justice, Institute of Medicine, National Research Council. 1999. Toward Environmental Justice: Research, Education, and Health Policy Needs.

Conca, Ken. 2005. *Governing Water: Contentious Transnational Politics and Global Institution Building.* Cambridge, MA: MIT Press.

Corcoran, Peter Blaze, and E. J. Arfen. 2007. *Higher Education and the Challenge of Sustainability: Problematics, Promise, and Practice.* New York: Springer.

Cristensen, Julia. 2008. *Big Box Reuse.* Cambridge, MA: MIT Press.

Culver, Keith, and David Castle, eds. 2008. *Aquaculture, Innovation, and Social Transformation.* New York: Springer.

Cunningham, Storm. 2002. *The Restoration Economy.* San Francisco: Berret-Koehler Publishers.

Curwell, S. R., Mark Deakin, and Martin Symes. 2005. *Sustainable Urban Development.* London, UK: Taylor and Francis.

Cutter, Susan L. 1993. *Living with Risk: The Geography of Technological Hazards.* London, New York: E. Arnold.

Cutter, Susan L. 2006. *Hazards, Vulnerability and Environmental Justice.* London: Sterling, VA: Earthscan.

Dauvergne, Peter. 2008. *The Shadows of Consumption: Consequences for the Global Environment.* Cambridge, MA: MIT Press.

Davenport, John, and Julia L. Davenport. 2006. *The Ecology of Transportation: Managing Mobility for the Environment.* New York: Springer.

Davidson, John H. 2002. "Agriculture." In *Stumbling toward Sustainability,* ed. John C. Dernbach, 351. Washington, DC: Environmental Law Institute.

Dernbach, John. ed. 1992. *Stumbling toward Sustainability.* Washington DC: Environmental Law Institute.

Devuyst, Dimitri et al. 2001. *How Green Is the City?: Sustainability Assessment and the Management of Urban Environments.* New York: Columbia University Press.

DiMento, Joseph F. C., and Pamela Doughman, eds. 2007. *Climate Change: What It Means for Us, Our Children, and Our Grandchildren.* Cambridge, MA: MIT Press.

Dixon, Tom et al., eds. 2007. *Sustainable Brownfields Regeneration: Livable Places from Problem Spaces.* Hoboken, NJ: Wiley-Blackwell.

Dowies, Mark. 2009. *Conservation Refugees: The Hundred-Year Conflict between Global Conservation and Native Peoples.* Cambridge, MA: MIT Press.

Duram, Leslie. 2005. *Good Growing: Why Organic Farming Works.* Lincoln: University of Nebraska Press.

Edwards, Andres R. 2005. *The Sustainability Revolution: Portrait of a Paradigm Shift.* Gabriola Island, BC: New Society Publishers.

Egan, Michael. 2009. *Barry Commoner and the Science of Survival.* Cambridge, MA: MIT Press.

Elkins, Paul. 2000. *Economic Growth and Environmental Sustainability.* London, UK: Routledge.

Emanuel, Kerry. 2007. *What We Know about Climate Change.* Cambridge, MA: MIT Press.

Gonenc, I. Ethem, 2004. *Coastal Lagoons: Ecosystem Processes and Modeling for Sustainable Use and Development.* Boca Raton, FL: CRC Press.

Fasulo, Linda. 2004. *An Insiders Guide to the UN.* New Haven, CT: Yale University Press.

Felleman, John. 1997. *Deep Information: The Role of Information Policy in Environmental Sustainability.* Westport, CT: Greenwood Press.

Folz, Richard C. et al., eds. 2003. *Islam and Ecology: A Bestowed Trust.* Cambridge, MA: Center for the Study of World Religions Harvard University Press.

Freidman, Thomas. 2008. *Hot, Flat, and Crowded: Why We Need a Green Revolution and How It Can Renew America.* New York: Farrar, Straus and Giroux.

Freyfogle, Eric T. 1993. *Justice and the Earth: Images for Our Planetary Survival.* New York: The Free Press.

Freyfogle, Eric T. 2003. *The Land We Share: Private Property and the Common Good.* Washington, DC: Island Press

Fuchs, D. A. 2003. *An Institutional Basis for Environmental Stewardship: The Structure and Quality of Property Rights.* New York: Springer.

Gabel, Medard, and Henry Bruner. 2003. *Global Inc.: An Atlas of the Multinational Corporation.* New York: New Press.

Geisler, Charles, and Gail Daneker, eds. 1997. *Property and Values: Alternatives to Public and Private Ownership.* Washington, DC: Island Press.

Gerrard, Michael B. 2007. *Global Climate Change and US Law.* Chicago: ABA Press.

Gibson, Robert B. et al. 2005. *Sustainability Assessment: Criteria and Process.* London, UK: Earthscan.

Gilbert, Charlene, and Eli Quinn. 2000. *Homecoming: The Story of African American Farmers.* Boston: Beacon Press.

Gore, Al. 1992. *Earth in the Balance: Ecology and the Human Spirit.* Boston: Houghton Mifflin.

Gore, Al. 2006. *An Inconvenient Truth: The Planetary Emergency of Global Warming and What We Can Do about It.* Emmaus, PA: Rodale Press.

Gore, Al. 2007. *The Assault on Reason.* New York: Penguin Press.

Gray, P. M. et al. "The Music of Nature and the Nature of Music" *Science* 291 (2001): 52–54.

Guerrant Edward O. Jr., Kayri Havens, and Mike Maunde, eds. 2004. *Ex Situ Plant Conservation Supporting Species Survival in the Wild.* Covelo, CA: Island Press.

Gunn, Angus M. 2003. *Unnatural Disasters: Case Studies of Human Induced Environmental Catastrophes.* Westport, CT: Greenwood Press.

Guthman, Julie. 2004. *Agrarian Dreams: The Paradox of Organic Farming in California.* San Francisco: University of California Press.

Hamilton, Clive. 2004. *Growth Fetish.* London, UK: Pluto Press.

Haroff, Kevin T., and Katherine Kirwan Moore. "Global Climate Change and the National Environmental Policy Act." *University of San Francisco Law Review* 42, no. 1 (2007): 155–83

Harper Bibles. 2008. *The Green Bible.* New York: HarperOne.

Hart, Stuart L. 2005. *Capitalism at the Crossroads: The Unlimited Business Opportunities in Solving the World's Most Difficult Problems.* Philadelphia: Wharton School Publishing.

Hawken, Paul. 1993. *The Ecology of Commerce: A Declaration of Sustainability.* New York: HarperBusiness.

Heal, Geoffrey. 2008. *When Principles Pay: Corporate Social Responsibility and the Bottom Line.* New York: Columbia University Business School Publishing.

Helming, Katharina, Marta Perez-Soba, and Paul Tabbush. 2008. *Sustainability Impact Assessment of Land Use Changes.* New York: Springer.

Hemmati, Minu et al. 2002. *Multi-Stakeholder Processes for Governance and Sustainability: Beyond Conflict and Deadlock.* London, UK: Earthscan.

Hess, David J. 2007. *Alternative Pathways in Science and Industry: Activism, Innovation, and the Environment in an Era of Globalization.* Cambridge, MA: MIT Press.

Hester, Randolph T. 2006. *Design for Ecological Democracy.* Cambridge, MA: MIT Press.

Hilty, Lorenz M. 2008. *Information Technology and Sustainability: Essays on the Relationship between Information Technology and Sustainability.* Gallen, Switzerland: Auflage.

Hitchcock, Darcy, and Marsha Willard. 2006. *The Business Guide to Sustainability: Practical Strategies and Tools for Organizations.* London, UK: Earthscan.

Hlavinek, Petr et al. 2009. *Risk Management of Water Supply and Sanitation Systems.* New York: Springer.

Hofrichter, Richard, ed. 2000. *Reclaiming the Environmental Debate: The Politics of Health in a Toxic Culture.* Cambridge, MA: MIT Press.

Holder, Jane. 2009. *Taking Stock of Environmental Assessment: Law, Policy, and Custom.* London, UK: RICS Books.

Hunkeler, David et al. 2008. *Environmental Life Cycle Costing.* Boca Raton, FL: CRC Press.

Josephson, Paul. 2004. *Resources Under Regimes: Technology, Environment, and the State.* Cambridge, MA: Harvard University Press.

Just, Richard E., and Sinaia Netanyahu. 1998. *Conflict and Cooperation on Trans-Boundary Water Resources.* New York: Springer.

Kasemir, Bernd, and Jill Jager. 2003. *Public Participation in Sustainability Science: A Handbook.* Cambridge, UK: Cambridge University Press.

Kassim, Tarek A., and Kenneth J. Williamson. 2005. *Environmental Impact Assessment of Recycled Wastes on Surface and Ground Waters.* New York: Springer.

King, Martin Luther, Jr. "Beyond Vietnam: Address Delivered to the Clergy and Laymen Concerned about Vietnam at Riverside Church, New York City April 4, 1967." In *A Call to Conscience [Sound Recording]: The Landmark Speeches of Dr. Martin Luther King, Jr.,* Clayborne Carson and Kris Shepard, eds. New York: Intellectual Properties Management, in association with Warner Books, 2001.

Klijn, Frans. 2007. *Ecosystem Classification for Environmental Management.* New York: Springer.

Kral, David M., ed. 1984. *Organic Farming: Current Technology and Its Role in Sustainable Agriculture.* Madison, WI: American Society of Agronomy.

Largent, Mark A., ed., George N. Vlahakis, Isabel Maria Malaquis, Nathan M. Brooks, Francois Regourd, Feza Gunergun, and David Wright. 2006. *Imperialism and Science: Social Impact and Interaction.* Santa Barbara, CA: ABC–CLIO.

Layzer, Judith A. 2008. *Natural Experiments: Ecosystem-Based Management and the Environment.* Cambridge, MA: MIT Press.

Leal Filho, Walter and Mario Salomone, eds. 2006. *Innovative Approaches to Education for Sustainable Development*. New York: Peter Lang.

Leal Filho, Walter. 2006. *Sustainability in the Australasian University Context*. New York: Peter Lang.

Leopold, Aldo. *A Sand County Almanac*. New York: Oxford University Press, 1949.

Levin, Henry M., and Patrick J. McEwan. 2000. *Cost Effectiveness Analysis: Methods and Applications*. Thousand Oaks, CA: Sage Publications.

Lumley, Sarah. 2002. *Sustainability and Degradation in Less Developed Countries: Immolating the Future?* Surrey, UK: Ashgate.

Lyson, Thomas, G. W. Steverson, and Rick Welsh, eds. 2008. *Food and the Mid-Level Farm: Renewing an Agriculture in the Middle*. Cambridge, MA: MIT Press.

Maida, Carl A. 2007. *Sustainability and Communities of Place*. New York: Berhahn Books.

Mander, Ulo, Hubert Wiggering and Katharina Helming. 2007. *Multifunctional Land Use*. New York: Springer.

Manjit, Kang S. 2007. *Agricultural and Environmental Sustainability: Considerations for the Future*. Boca Raton, FL: CRC Press.

McConnel, Grant. 1972. "The Failures and Success of Organized Conservation." In *Environment and Americans: The Problem of Priorities*. Roderick Nash, ed. Huntington, NY: R. E. Krieger.

McIntosh, Roderick J., Joseph A. Tainter, and Susan Keech Mcintosh. 2000. *The Way the Wind Blows: Climate Change, History, and Human Action*. New York: Columbia University Press.

McNeill, J. R. 2000. *Something New under the Sun: An Environmental History of the Twentieth Century World*. New York: W. W. Norton.

Meyers, Norman, and Jennifer Kent. 2001. *Perverse Subsidies: How Tax Dollars Harm the Environment and the Economy*. Washington, DC: Island Press.

Mission, C. F. 2005. *Sharing God's Planet: A Christian Vision for a Sustainable Future*. London, UK: Church House Publishing.

Mitchell, Ronald B., William C. Clark, David W. Cash, and Nancy M. Dickson. 2006. *Global Environmental Assessments: Information and Influence*. Cambridge, MA: MIT Press.

Montgomery, David R. 2007. *Dirt: The Erosion of Civilizations*. San Francisco: University of California Press.

Monto, Mani, L. S. Ganesh, and Koshy Varghese. 2005. *Sustainability and Human Settlements: Fundamental Issues, Modeling and Simulations*. Thousand Oaks, CA: Sage Publications.

Moran, Emilio F., and Elinor Ostrom. 2005. *Seeing the Forest and the Trees: Human-Environment Interactions in Forest Ecosystems*. Cambridge, MA: MIT Press.

Morello-Frosch, Rachel, Manuel Pastor, and James Saad. "EJ and Southern California's Riskscape: The Distribution of Air Toxics Exposures and Health Risks among Diverse Communities." *Urban Affairs Review* 36 (2001): 551.

Mowforth, Martin, and Ian Munt. 1998. *Tourism and Sustainability: New Tourism in the Third World*. Andover, UK: Routledge.

Mulamoottil, George, Edward A. McBean, and Frank Rovers. 1998. *Constructed Wetlands for the Treatment of Landfill Leachates*. Boca Raton, FL: CRC Press.

Myers, Nancy J., and Carolyn Raffensperger, eds. 2005. *Precautionary Tools for Reshaping Environmental Policy*. Cambridge, MA: MIT Press.

National Academies Press. 2004. *Endangered and Threatened Fishes in the Klamath River Basin: Causes of Decline and Strategies for Recovery*. Washington: DC: National Academies Press.

Nicholsen, Shierry Weber. 2003. *The Love of Nature and the End of the World: The Unspoken Dimension of Environmental Concern*. Cambridge, MA: MIT Press.

Ohshita, Stephanie B. "The Scientific and International Context for Climate Change Initiatives." *University of San Francisco Law Review* 42, no. 1 (2007): 1–37.

O'Riordan, Timothy, and Susanne Stoll-Kleemann, eds. 2002. *Biodiversity, Sustainability and Human Communities: Protecting Beyond the Protected*. Cambridge, UK: Cambridge University Press.

Paarlberg, Robert. 2009. *Starved for Science: How Biotechnology Is Being Kept out of Africa*. Cambridge, MA: Center for the Study of World Religions, Harvard University Press.

Parr, Adrian. 2009. *Hijacking Sustainability*. Cambridge, MA: MIT Press.

Parry, M. L., O. F. Canziani, et al., eds. 2007. *Climate Change 2007: Impacts, Adaptation and Vulnerability. Contribution of Working Group II to the Fourth Assessment Report of the Intergovernmental Panel on Climate Change*. Cambridge, UK: Cambridge University Press.

Pelling, Mark. 2003. *The Vulnerability of Cities: Natural Disasters and Social Resilience*. London, UK: Earthscan.

Pirages, Dennis, and Ken Cousins. 2005. *From Resource Scarcity to Ecological Security: Exploring New Limits to Growth.* Cambridge, MA: MIT Press.

Platt, Rutherford H. 1996. *Land Use and Society: Geography, Law, and Public Policy.* Washington, DC: Island Press.

Polanyi, Karl. 1944. *The Great Transformation.* Boston: Beacon Press.

Porter, Douglas. 2002. *Making Smart Growth Work.* Washington, DC: Urban Land Institute.

Pozzo, Barbara, ed. 2007. *Property and Environment: Old and New Remedies to Protect Natural Resources in the European Context.* Durham, NC: Carolina Academic Press.

Princen, Thomas et al., eds. 2002. *Confronting Consumption.* Cambridge, MA: MIT Press.

Rappaport, Ann, and Sarah Hammond Creighton. 2007. *Degrees That Matter: Climate Change and the University.* Cambridge, MA: MIT Press.

Rappaport, Roy A. 1994. "Human Environment and the Notion of Impact." In *Who Pays the Price? The Sociocultural Context of Environmental Crisis,* ed. Barbara Rose Johnston. Washington, DC: Island Press.

Raymond, Leigh Stafford. 2003. *Private Rights in Public Resources: Equity and Property Allocation in Market-Based Environmental Policy.* Washington, DC: Resources for the Future.

Robinson, John, and Elizabeth Bennett. 1999. *Hunting for Sustainability in Tropical Forests.* New York: Columbia University Press.

Robson, Mark G., and William E. Toscano. 2007. *Risk Assessment for Environmental Health.* Hoboken, NJ: Jossey-Bass.

Rogers, Heather. 2006. *Gone Tomorrow: The Hidden Life of Garbage.* New York: New Press.

Roodman, David Malin. 1996. *Paying the Piper: Subsidies, Politics, and the Environment.* Washington, DC: Worldwatch Institute.

Roseland, Mark. 2005. *Toward Sustainable Communities: Resources for Citizens and Their Governments.* Gabriola Island, BC: New Society Publishers.

Rosemarin, Arno et al. 2008. *Pathways for Sustainable Sanitation: Achieving Millennium Development Goals.* London, UK: IWA.

Roser, Dominik et al. 2008. *Sustainable Use of Forest Biomass for Energy.* New York: Springer.

Roszak, Theodore, Mary E. Gomes, and Allen D. Kanner, eds. 1995. *Ecopsychology: Restoring the Earth, Healing the Mind.* San Francisco: Sierra Club Books.

Sabatier, Paul et al., eds. 2005. *Swimming Upstream: Collaborative Approaches to Watershed Management.* Cambridge, MA: MIT Press.

Sagoff, Mark. 1988. *The Economy of the Earth: Philosophy, Law, and the Environment.* New York: Cambridge University Press.

Schaltegger, Stefan et al. 2006. *Sustainability Accounting and Reporting.* New York: Springer.

Shiva, Vandana. 2001. *Water Wars: Pollution, Profits, and Privatization.* Lexington, MA: South End Press.

Slobodchikoff, C. N. 2002. "Cognition and Communication in Prairie Dogs," In *The Cognitive Animal: Empirical and Theoretical Perspectives on Animal Cognition.* ed. Marc Bekoff, Colin Allen, and Gordon M. Burghard. Cambridge, MA: MIT Press.

Smil, Vaclav. 2008. *The Earth's Biosphere: Evolution, Dynamics, and Change.* Cambridge, MA: MIT Press.

Spellman, Frank R. 2007. *Environmental Management of Concentrated Animal Feeding Operations.* Boca Raton, FL: CRC Press.

Stein, Bruce A. et al., eds. 2000. *Our Precious Heritage: The Status of Biodiversity in the United States.* New York: Oxford University Press.

Stern, Alissa J. 2000. *The Process of Business/Environmental Collaborations: Partnering for Sustainability.* Westport, CT: Greenwood Publishing.

Thomashow, Mitchell. 2003. *Bringing the Biosphere Home: Learning How to Perceive Global Environmental Change.* Cambridge, MA: MIT Press.

Thoreau, Henry David. 1854. *Walden: Life in the Woods.* Boston: Tickner and Fields.

Tilbury, Daniella, and David Wortman. 2004. *Engaging People in Sustainability.* Gland, Switzerland: International Union for Conservation of Nature and Natural Resources.

Tocqueville, Alexis de. 1863. *Democracy in America.* Ann Arbor: University of Michigan.

Tow, Philip, Ian Cooper, and Ian Partridge. 2009. *Rainfed Farming Systems.* New York: Springer.

United Nations. 2004. *Basic Facts about the United Nations.* New York: United Nations.

Volk, Tyler. 2008. *CO2 Rising: The World's Greatest Environmental Challenge.* Cambridge, MA: MIT Press.

Warner, Keith Douglass. 2007. *Agroecology in Action: Extending Alternative Agriculture through Social Networks.* Cambridge, MA: MIT Press.

Webster, D. G. 2008. *Adaptive Governance: The Dynamics of Atlantic Fisheries Management.* Cambridge, MA: MIT Press.

Weiss, Charles, and William B. Bonvillian. 2009. *Structuring an Energy Technology Revolution.* Cambridge, MA: MIT Press.

Whiteside, Kerry H. 2006. *Precautionary Politics: Principle and Practice in Confronting Environmental Risk.* Cambridge, MA: MIT Press.

Wildman, Stephanie M. 1996. *Privilege Revealed: How Invisible Preference Undermines America.* New York: New York University Press.

Williams, Juan. 2006. *Black Farmers in America.* Lexington: University of Kentucky Press.

Wilson, Edward O. 1984. *Biophilia.* Cambridge, MA: Harvard University Press.

World Commission on Environment and Development. 1987. *Our Common Future: Report of the World Commission on Environment and Development.* New York: Oxford University Press.

Yanful, Earnest K. 2009. *Appropriate Technologies for Environmental Protection in the Developing World.* Dordrecht, Netherlands: Springer Press.

Zovanyi, Gabor. *Growth Management for a Sustainable Future: Ecological Sustainability as the New Growth Management for the 21st Century.* Westport, CT: Greenwood Publishing.

INDEX

ABOUT THE AUTHORS

ROBIN MORRIS COLLIN, professor of law, Willamette College of Law. Professor Morris Collin has taught at McGeorge School of Law, Tulane School of Law, Pepperdine Law School, Washington and Lee School of Law, University of Oregon School of Law, and Willamette School of Law. She has numerous publications in the area of sustainability and holds the David Brower Lifetime Achievement Award. In April 2009, she was awarded the *Judith Ramaley Faculty Award for Civic Engagement in Sustainability,* Oregon Campus Compact. Professor Collin was the first law professor to teach sustainability in the United States in 1993 and has taught it ever since. She has served as an advisor to state and federal environmental agencies. She has also litigated court cases and provided legislative testimony on many important environmental issues. She is currently working with the Oregon State Bar Association to find ways to integrate sustainability into legal practice. She is also appointed to the Oregon Environmental Justice Advisory Group.

ROBERT WILLIAM COLLIN is the senior research scholar at the Center for Sustainable Communities at Willamette University. He has been a professor of law, planning, and of social work, teaching at the University of Auckland, New Zealand; the University of Virginia Department of Urban and Environmental Studies; the University of Oregon's Department of Environmental Studies; Cleveland State University Department of Social Work; Jackson State University Department of Urban and Regional Planning; Lewis and Clark College of Law; and Willamette University College of Law. He has published many articles, book chapters, and book reviews. He has served as an advisor to state and federal environmental agencies and currently serves as chair of the Oregon Environmental Justice Advisory Group. His last two books are *The US Environmental Protection Agency: Cleaning up America's Act,* and *Battleground: Environment* (2 volumes).